GEOGRAPHERS
Biobibliographical Studies

VOLUME 18

GEOGRAPHERS: BIOBIBLIOGRAPHICAL STUDIES

GEOGRAPHERS
Biobibliographical
Studies

VOLUME 18

Edited by Patrick H. Armstrong
and Geoffrey J. Martin

on behalf of the
Commission on the History of Geographical Thought
of the International Geographical Union and the
International Union of the History and Philosophy of Science

MANSELL
London and Washington

First published 1998 by
Mansell Publishing Limited, *A Cassell imprint*
Wellington House, 125 Strand, London WC2R 0BB, England
PO Box 605, Herndon, Virginia 20172, USA

British Library Cataloguing in Publication Data

Geographers: biobibliographical studies – Vol. 18
1. Geographers – Biography – Periodicals
910′92′2 G67

ISBN 0–7201–2339–9

Typeset by BookEns Ltd, Royston, Herts.
Printed and bound in Great Britain by Creative Print and Design (Wales), Ebbw Vale

Contents

Introduction

The biobibliographical sortie into the world of geography, represented by this series, now reaches its eighteenth publication. Some two hundred and eighty geographers have now been presented in similar format over the last twenty years. We recognize in that format the headings: education, life and work; scientific ideas and geographical thought; influence and spread of ideas; and, of course, the valuable end matter. While this may restrict and confine somewhat, it also offers a parallel contract by which the reader may attempt comparative assessment of individual opportunity and contribution. Of course, contributions made in the fifteenth century, for example, will invariably be of a less sophisticated nature than those made later. The development of a field of learning which we identify as geography has been in the making for a long time, arguably since the first hominids appeared upon this earth. The role of learning has increased as the millennia have passed. Learning begets learning, and we benefit greatly from those who have gone before – T.C. Chamberlin referred to this as the staircase principle.

In this advance of knowledge, the geographer (under different labels) has played a most significant role, and that is where our biobibliographical series makes its contribution.

In this volume, the Portuguese explorer Ferdinand Magellan (c.1480–1521) is offered as the earliest of our subjects. He extended European knowledge of the world that was beyond the boundaries of Europe – knowledge of the Pacific, the Philippines, the Strait (of Magellan), the Spice Islands and, indeed, of Magellanic Clouds. He brought Spanish culture, economic system and catholicism to the Philippines, which land functioned as a hearth to diffusion.

David Thompson (1770–1857) was another of those extraordinary explorers who helped reveal and make known a part of this planetary surface, a part which all would now think of as western Canada. Thompson, we are informed, travelled 50,000 miles by foot, horseback or canoe, carrying out surveys which led to the first adequate map of the Canadian west. Apart from surveying and mapping the Canada–USA boundary line, he also was first to accurately locate and record the sources of the Mississippi and Missouri rivers. The narrative which he produced in his journals and in his other writings constitutes a most excellent regional geography of a part of the earth then so little known.

Another person who spent most of his life in Canada was Frank Gilbert Roe (1878–1973). A hardy soul, who lived the outdoor life as a farmer and railroadman, he produced his first publication at the age of 51, and his main work, *The North American Buffalo*, when he was 73. His writing came from a very personal experience leavened with an intellectual curiosity and an Antaean quality of being. This is revealed in an autobiography that treats the period 1894–1944.

John Lort Stokes (1811–85), surveyor and naval officer in the British Admiralty's Hydrographic Service, contributed substantially to the accuracy of coastal charts and inland exploratory surveys of Australia. His activities extended elsewhere, more notably to New Zealand, the Cape of Good Hope and the coast of South America. His association with scientists, including Charles Darwin, provided him with both facility and encouragement to continue his work. As late as the Second World War, a number of his Australian charts were still in use.

These four men lived their lives and made their contributions in a time before the advent of geographical discipline, that late nineteenth-century (but varying from nation to nation) Rubicon which has received so little attention, yet which transformed a congeries of subject matter (the stuff of exploration and survey) into a coherent whole from which further intellectual growth is possible. The remaining seven figures advanced in this publication are drawn from the post-disciplinal stage of intellection, and therefore are a part of the developing saga of what we call modern geography.

The first of these 'modern' geographers, Frederic Clements (1874–1945), was essentially an ecologist before that field (or term) was recognized in North America. He was a founding member of the Association of American Geographers and one of several ecologists to share with geographers, and especially biogeographers, some ideas including an extension and analysis of plant succession and the interaction of succession and physiography. His work was continually caught up with geomorphology and microclimatology, but his elaboration of vegetation types in North America meant that he contributed more to geography than he received from it.

Two other American climatologists, Charles F. Brooks (1891–1958) and C. Warren Thornthwaite (1899–1963), were both interested in plants as indicators; this was especially true of Thornthwaite. Brooks worked intermittently with the Bureau of Farm Management in the US Department of Agriculture. He also worked as an academic affiliated with Yale, Clark and Harvard universities. His climatology was placed squarely in the geographical tradition and was concerned to a large extent with its human impact. This interest may have derived in part from his mentor, Robert DeCourcy Ward, himself much interested in physiological climatology. Thornthwaite, also a climatologist, was very interested in plants. He established a laboratory at Seabrook Farms, where his research was put to good use to help with crop harvests. He was also responsible for a new and rational classification of climate (published in 1948) and the development of a concept of potential evapotranspiration. He argued strongly for the development of a climatology, grounded in the sciences, distinct from the study of formal geography.

Kenneth J. Mason (1887–1976) and Roger Dion (1896–1981), two Europeans essentially of the twentieth century, also win inclusion. Mason, who took the first chair in the department of geography at Oxford University in 1932 at the age of 45 and without formal experience or qualifications for university life, had made substantial contributions to survey and exploration in Kashmir and the Karakorams. Much of this work was undertaken as an employee of the Survey of India. He became particularly interested in high mountain environments, ice, related hazards and exploration, and wrote about them, perhaps favouring publication in *The Geographical Journal*. He was a firm and resolute figure at Oxford and led the department through a difficult period. Terence Edward Armstrong (1920–96) was also an explorer of ice environments and a meticulous worker, author and authority on such areas.

Roger Dion, known earlier in his career for his contribution to regional and rural geographies, later came to specialize more in the historical geography of earlier times – from the Greco-Roman world to the Middle Ages. He left behind an impressive bibliography of works which are in part in the tradition of Vidal de la Blache and which are evergreen. His later work in historical geography encouraged renewed effort in that direction by younger French workers, beginning perhaps in the 1980s.

Finally, there stands revealed the contribution of Shiba P. Chatterjee (1903–89), the first Indian (and Asian) to be elected president of the International Geographical Union. He studied in Europe, became a very capable geomorphologist, taught a variety of other courses, and founded a department at the University of Calcutta which included a cartographic unit that was responsible for the Atlas of India. He worked in most difficult circumstances throughout his life, but left behind an exemplary legacy.

The co-editors would like to take this opportunity to recognize the work – above and beyond the call of duty – of Mrs Joan Boughton (New Haven, Connecticut), in the preparation of the manuscript for the publisher.

Patrick H. Armstrong and Geoffrey J. Martin

List of Abbreviations

AAG	Association of American Geographers
AMS	American Meteorological Society
CCL	Commission for Climatology
GPS	Global Positioning System
IGU	International Geographical Union
IMO	International Meteorological Organization
MIT	Massachusetts Institute of Technology
NAO	National Atlas Organization
NATMO	National Atlas and Thematic Mapping Organization
NID	Naval Intelligence Division
SPRI	Scott Polar Research Institute
WMO	World Meteorological Organization

Terence Edward Armstrong

1920–1996

Peter Speak

Terence Edward Armstrong died at his home, Harston House, Harston, South Cambridgeshire, on 21 February 1996. The whole of his professional life (36 years) was spent at the Scott Polar Research Institute (SPRI) in Cambridge, where he established for himself and for the SPRI an international reputation in polar research and research on the Russian Arctic. He was regarded throughout the UK as the foremost authority on the Russian north: its peoples and their environment, and its economic development. He was successively Research Fellow, Assistant Director of Research, Reader in Arctic Studies and Acting Director of the SPRI. Whilst his introduction to the polar north came through his knowledge of Russian, he also became an expert on the circumpolar north in general, and was well known for his contributions to studies of Canada, Alaska and Greenland. During the time that Dr Gordon Robin was director of the SPRI from 1958 to 1983, the research and teaching was largely divided between them: Gordon Robin concentrated on the Antarctic and the physical sciences, and Terence on the Arctic and the social sciences. Harry King had joined the SPRI as Librarian in 1955, and Brian Roberts in 1946 as part-time Research Fellow: they became a very powerful 'group of four' which lasted until Brian's death in 1978 and the retirement of the others in 1982–3. Together they created an institute with a reputation internationally as a centre of excellence.

Apart from his linguistic abilities, Terence Armstrong became known as an economic and social geographer, yet he was never trained as one: 'I never learned Geography as a single subject, but once appointed to SPRI I read all the standard textbooks in the subject.' He had a natural appreciation of the problems of native peoples and environment and travelled widely across all parts of the Arctic. In an interview at the SPRI recorded some months before his death, he claimed that his career was a series of chance events: 'I was not driven by a boyish desire to find out about the Poles, but, like many others came to the subject by chance.' He was elected a Fellow of the Royal Geographical Society and often addressed its meetings. The society gave him the Cuthbert Peek Award in 1954, and the Victoria Medal in 1978. He made an important contribution to academic life in Cambridge, helping in the initial stages of the founding of Clare Hall, one of the post-war graduate colleges, becoming a Founder Fellow in 1964, the first senior tutor in 1966 and vice-president in the 1980s. Former president of the college, Sir Michael Stoker, has written, 'I never heard him speak ill of anyone – I suspect that during nearly three decades Terence was the most popular member of Clare Hall. Everyone loved him.' Indeed, all who knew him admired his fine scholarship, his immense patience and tolerance of other people's points of view, his great kindness to students and colleagues alike, and his sense of fun and infectious laughter.

The Armstrong family came originally from northern England, close to the Scottish Borders, where they were sheep farmers, but during the plantation period were encouraged to move to Northern Ireland, where they became linen merchants. Although both of Armstrong's parents were from Ulster, they had moved to England before any of their six children were born. Armstrong was born on 7 April 1920 at Oxted in Surrey and attended Winchester School. It was whilst he was there, in the upper school, that a distinguished Russian historian, Sir Bernard Pares, gave a talk in 1937 encouraging the boys to study the Russian language. One of the masters agreed to learn Russian and to try to keep two steps ahead of the class: Armstrong was one of the first to accept the challenge.

1. Education, Life and Work

Armstrong went up to Magdalene College, Cambridge, in 1938 to read modern languages – French and German. However, he was soon discovered by the legendary Russian scholar and teacher Dr (later Dame) Elizabeth Hill. She had heard of his Winchester training and persuaded him to join the small Department of Slavonic Studies under her tuition, to continue with Russian and to drop German. Armstrong took first class honours in French and Russian in Part One of the Tripos and then in 1940 joined the Army Intelligence Corps. When asked whether he was able to use his Russian in the forces, he replied, with a characteristic smile, 'Oh, once the Army knew that I could speak Russian they immediately dispatched me to North Africa!'. He served also in Italy, in Holland, where he parachuted at Arnhem behind enemy lines and was wounded but made his way back to allied lines south of the Rhine, and in Norway, where he was amongst the first in the liberation of Oslo.

In 1946, after demobilization, he returned to Cambridge and sought the advice of Elizabeth Hill on his future. She had learned that the SPRI was thinking of appointing a Russian language specialist and advised him to apply. So, with little knowledge of the polar world, Armstrong approached James (later Sir James) Wordie, at that time chairman of the SPRI's Committee of Management and later master of St John's College, and applied for the post of Research Fellow in Russian Studies. There was no formal interviewing board: Armstrong saw Wordie in his study in St John's and later met Launcelot Fleming, at that time director of the SPRI, and Brian Roberts, the SPRI's 'guru', whose idea it was to have a Russian specialist on the staff. Armstrong was appointed to the Fellowship, which was funded partly from a Treasury grant, and partly from rent money accumulated during wartime, when the SPRI was home to senior geographers employed by the Admiralty to write a series of naval intelligence handbooks. The Fellowship lasted from 1947 to 1956, when Armstrong was appointed Assistant Director of Research; he succeeded in 1977 to an *ad hominem* Readership in Arctic Studies. In 1947 the SPRI was still under the guiding hand of its first director, Frank Debenham, who was also the first Professor of Geography in the university. Armstrong realized that his work had to meet geographical criteria, but he needed to select, for a doctoral thesis, a subject that required language skills in Russian but without any other specialist training. His choice was *The Development of the Northern Sea Route*, an examination of the history, economic development and commercial potential of the north-east passage across the Russian Arctic Ocean. No-one else was working on this topic in English at that time, and Armstrong became the British specialist. His PhD was awarded in 1951 and this led him to further studies of the economic effects of sea ice on northern navigation and sponsorship by the Royal Navy Scientific Survey. His doctoral thesis provided the title of his first book, published in 1952: *The Northern Sea Route: Soviet Exploration of the North East Passage.* He maintained his interest in the northern sea route and published an annual report in the SPRI's house journal, *Polar Record*, of the volume of shipping, the sizes and names of ships, and the growth of the Soviet ice-breaker fleet from the information that he gleaned from his daily reading of *Pravda* and from the periodical *Vodnyi transport.* He stated in 1994: 'It was only part of the truth, but Russians tell me now that it was mainly correct.' Since perestroika, there has been renewed interest in the possibilities of increased utilization of the passage by Western nations, and in 1994 Armstrong became the British representative on an International Advisory Committee on the Northern Sea Route to gather data and conduct research on this problem. Only a few days before his death he had completed an historical account of the northern sea route for the committee and had published with Lawson Brigham, a former ice-breaker captain of the US Coast Guard Service and the US representative on the committee, a review of present practices ('The Northern Sea Route 1995', *Polar Record*, Vol. 32, no. 183 (1996), 353–5).

His book *The Northern Sea Route* was followed by *The Russians in the Arctic* (1960), and *Russian Settlement in the North* (1965). Although travel to the Soviet Union was difficult in the days of Stalin, Armstrong managed eventually to visit (from 1956 onwards) many parts of the country, not only in the Siberian north but many other parts as well, including research centres in Leningrad and Moscow, and to arrange exchanges with SPRI personnel and to forge lasting friendships with Russian scholars. These included Academician Treshnikov, who had been in charge of one of the early Russian ice-stations and later became director of the Arctic Research Institute of Leningrad (now the Arctic and Antarctic Research Institute of St Petersburg), Professor Grigori A. Agranat of the Russian Academy of Sciences, Moscow, Admiral Vasilii F. Burkhanov, head of the Northern Sea Route, Innokenti P. Gerasimov and Vladimir M. Kotlyakov, directors of the Institute of Geography of the USSR Academy of Sciences. Agranat, in a recent appreciation of Armstrong published in *Polar Record* (vol. 32, no. 182, 265–70 (1996)), has written of Armstrong's friendship with Boris Kremer, a famous Soviet *polyarnik* (polar explorer), who stated:

> In the Soviet Union, we greatly value Terence's fundamental works on the history of the development and exploration of the Soviet Arctic and north, the development of the Northern Sea Route, and issues relating to the northern indigenous people. In Russia, his works are especially esteemed for their objectivity. Even though there was socialism and capitalism, what Terence did was tell the truth. In his work he spoke the truth notwithstanding the ideology.

The Russian north was not the only *leitmotif* in Armstrong's scholarship. In July

1953 he flew over the North Pole from Thule, northern Greenland, by courtesy of the RAF Flying College, Manby, in the course of its high-latitude navigational exercises, and in 1954 he was invited to join the maiden voyage of HMCS *Labrador* through the ice-packed Northwest Passage. The voyage was ostensibly to record ice conditions along the route, but was also a Canadian gesture to the USA over sovereignty rights through the straits and seas of the American Arctic. Armstrong persuaded Captain Robertson to land him by helicopter on the most westerly tip of Alaska, Prince of Wales Point on the Seward Peninsula, so that he could make his way home across the continent. As there were no immigration officials at Prince of Wales Point, he was technically an illegal entrant to the USA, and had difficulty with officials when attempting to leave from Boston; only after the passport controller had telephoned Iris, Armstrong's wife in Harston, Cambridgeshire, was his story believed!

His contract with the Canadian Defence Board on the reporting of sea ice conditions and similar work carried out for the Royal Navy Scientific Survey led to his devising a set of symbols to show sea ice characteristics and the publication in 1958 of an ice atlas, *Atlas of Sea Ice North of the USSR*. The symbols were later used within the institute by Charles Swithinbank and John Heap to produce working ice charts of both Arctic and Antarctic waters. In this way, Armstrong made an early contribution to sea ice studies which form such an important part of the work of the SPRI today. The photographs taken when he was aboard HMCS *Labrador* have formed the basis of an important collection of sea ice images held in the SPRI and were used in the important *Illustrated Glossary of Snow and Ice* published in 1966 in collaboration with Charles Swithinbank and Brian Roberts. Although essentially a social scientist, Armstrong continued his interests in glaciology by serving as treasurer, from 1965 to 1970, of the British (now International) Glaciological Society, which had its secretariat within the SPRI.

After his voyage on the *Labrador*, Armstrong became a frequent traveller in Canada and Alaska and visited virtually all the centres of Arctic and northern studies; he made friends wherever he went and invited polar scholars back to Cambridge to the mutual enrichment of the work of the SPRI and similar institutions in North America. Three of these scholars were especially important and all became life-long friends and frequent visitors to the SPRI and to Harston House: Graham Rowley, science advisor to the Department of Indian Affairs and Northern Development based in Ottawa; George Rogers, the principal economic advisor to the State of Alaska; and Frank Darnell, a distinguished educator of northern peoples with a life-long interest in minority peoples around the world. It is not surprising to those who knew him well that shortly after the announcement of his death a memorial service was held in the University of Alaska attended by former students and colleagues.

Another friendship which lasted almost fifty years was with Bertrand Imbert, sometime leader of French Antarctic expeditions to Terre Adélie. They met at the SPRI when Imbert was planning the first French Antarctic expedition and came to the SPRI for information. They had planned joint authorship of the book *The Russians in the Arctic* but Imbert had not supplied any copy by the time the draft was ready. Armstrong, with characteristic wit, dedicated the book 'To my silent partner'.

In 1970 Armstrong was awarded a scholarship by the National Science Foundation of America to contribute to the US Programme 'Man and the Arctic'. He enjoyed a 'half-sabbatical' at the Fairbanks campus of the University of Alaska in Vic Fischer's Department of Economics and Commerce. Armstrong became interested at this time in the education of Indians and Eskimos. This interest led to a series of seminars on education that has lasted until the present day. A group of

Alaskans, Canadians, Greenlanders and Scandinavians (most of whom were of indigenous stock) approached Armstrong and Frank Darnell in the mid-1970s to join in an effort to improve the education of indigenous peoples in the Arctic and sub-Arctic. The group became known as 'The Steering Committee for a Series of International Seminars on Cross-Cultural Education in the Circumpolar North'. The membership included leading pedagogic figures from Alaska, Canada, Greenland, the Nordic Saami Institute, and other Inuit and Métis organizations. Armstrong managed the administration of the group from the SPRI, initially with the aid of a Ford Foundation grant. The early meetings were held in Cambridge and then in many centres across the far north. It is not surprising that the most influential general book about northern affairs, *The Circumpolar North* (1978), came from the combined authorship of Armstrong, George Rogers and Graham Rowley, or that the University of Alaska, Fairbanks should award him in 1980 the honorary degree of DSc. He had, in 1963, been awarded an honorary LLD by McGill University, Montreal.

In 1975, Gordon Robin and Armstrong initiated at the SPRI the Diploma in Polar Studies, a course lasting nominally one year (in fact from October to July), in order to bring a formalized teaching function to the SPRI. From its foundation in 1920, the SPRI had remained purely a research institute, although there were supervisory commitments to postgraduate students. There was no real or perceived demand initially for the Diploma course and very little publicity, but along those polar communication links built up over many years by members of the SPRI excellent students were forthcoming, many of whom were veterans of the polar world, coming from both hemispheres. Armstrong not only acted as director of the course from time to time but also persuaded friends and acquaintances of the highest authority to come and give seminars. The success of the diploma led in 1980 to its upgrading to the present MPhil in Polar Studies and to its widespread reputation for excellence. It was the first such course in the world; now other centres both within polar regions and outside them have adopted comparable courses. Armstrong always contributed to the seminar programme (even long after he had officially retired) with topics on the Russian Arctic, the northern sea route, northern native peoples, etc. Particularly memorable was the seminar he gave on the Siberian journeys of Vitus Bering on behalf of Peter the Great. It is unlikely that any other historian outside the Russian Federation could speak with such authority, or with greater humour. He was especially delighted when the first students from Russia arrived to take the course in 1994, one from St Petersburg and one from Yakutsk.

2. Scientific Ideas and Geographical Thought

Richard Bridge, a former research student, has written: 'Terence was, to his credit, not an intellectual's intellectual, he was not a constructor of abstract theories. He was a pragmatic idealist.' Whilst not trained in the first place as a geographer, he had a natural affinity for the subject, regarding himself as a human geographer and being seen by others in this way. Like most geographers fifty years ago, his interests were eclectic, lying within the social sciences, and he viewed northern studies holistically; the truth should be sought not through one academic discipline alone, but through the combined efforts of historian, economist, political scientist and geographer. He was concerned to interpret northern relationships by seeking the views of the indigenous peoples through their own language, not only through

Russian and other Asiatic and Scandinavian languages, but throughout the whole of the circumpolar north, where orthographies were little known and often difficult to interpret. As far as he could, he visited all parts of the north, sat down with indigenous peoples and discussed with them their histories and aspirations. His work was restricted wholly to the Arctic and sub-Arctic, for he never visited Antarctica, where academic problems were primarily (but not exclusively) those of the physical sciences. Whilst he was attracted to environmentalism as a key to the understanding of northern geography, he fully appreciated that cold, latitude, insolation and inaccessibility were not necessarily such exacting controls of development today as in past times. There were few who knew better than Armstrong the fundamental differences between the economic development of the Russian north and the American north, the former controlled by an autocratic ideology and the latter by market forces. He maintained an objective and dispassionate stance towards Russian literature and was critical of political motivation, particularly over ecological disasters, but was not afraid to praise when he thought it right to do so. His books on the Russian north were not reviewed in the Russian press. 'Actually I regarded it as rather a compliment,' he said, 'for if they had been the reviewers might have been tempted to toe the Party line and have been obliged to say things they did not believe.'

In an important paper, 'Ethical problems of Northern development', published in *Polar Record* (Vol. 19 (1978), 118), Armstrong discussed the effects of 'development', interpreted as 'an incursion of the south into the north', and raised the question: 'is such activity in essence a plundering of local resources or amenities, carried out by a distant authority for its own enrichment or protection, or is it reasonable use of that authority's sovereign territory?' In the conclusion to the paper he offers 'some guidelines':

> ensure that the legislative and administrative structure permits the local inhabitants a reasonable voice in any decision affecting the development of their area; development should be minimally disruptive of the physical and social environment; rapid fluctuations are to be avoided; a royalty on all mineral exploitation [should] be paid to the local community.

He was frequently consulted by native peoples' organizations, as well as by state and federal governments.

Armstrong always maintained that the two people who had the greatest influence on his life were the late Dame Elizabeth Hill (his acclaimed fluency in the Russian language), and Dr Brian Roberts, his senior colleague at the SPRI until his death in 1978. In a recently published *Festschrift* for Brian Roberts (*Polar Pundit*, H. G. R. King and Ann Savours), Armstrong contributed the text of the oration he gave at the memorial service for Brian Roberts in St Paul's Cathedral. It included the following: 'Another attitude characteristic of Brian was the way in which he placed the good of mankind and the world of nature above sectional interests.' These words could well be used also of Terence Armstrong.

3. Influence and Spread of Ideas

Without question, a major contribution that Armstrong made to scholarship was his analysis of Russian texts and his translation of many of them into English. In his early professional years, information on the Russian north was extremely difficult to acquire. On all his visits to the former Soviet Union since 1956, he endeavoured to bring home texts not available in the West and to establish exchanges of literature. He visited the scientific periodicals library in Novosobirsk on one of his early visits and was given the freedom of the shelves. Collecting as many books and journals as he could carry back to Cambridge, he left behind a trolley full of other selected works, not expecting that he would see them again. But over the succeeding months, they were all delivered to the SPRI. 'Because of the non-convertibility of the rouble to western currencies I always insisted on an exchange, book-for-book, and not, value-for-value.' In these days before the SPRI had appointed a Russian cataloguer, he set aside one day each week to keep abreast of the translation and indexing of Russian materials. In this way, the SPRI built up an unrivalled collection of books, journals, pamphlets and statistical data on the Russian north to the extent that research scholars from Russia have been known to visit the SPRI, where their needs were more readily supplied than in Russia itself. He made a major contribution to the SPRI's own journal, *Polar Record*, the oldest journal in the world devoted to polar regions. For over forty years there was hardly a single issue of this quarterly that did not contain a contribution from him.

Armstrong also made an important contribution to the Hakluyt Society, which is committed to the publication of scholarly editions of records of voyages and other geographical material of the past. He joined the society in 1960 and became joint honorary secretary with the late Dr Eila Campbell, Professor of Geography at Birkbeck College, London, in 1966. He remained joint secretary for twenty-five years and saw more than fifty publications through the press, including his edited version of *Yermak's Campaign in Siberia* (1975).

John Heap, present director of the SPRI, has written: 'Terence Armstrong did not aspire to leadership in the hierarchical sense, but whatever he put his hand to was changed for the better' (*The Independent*, 1 March 1996). Nor was Armstrong the polar scholar turned explorer: he was much too gregarious to become a member of a small, drifting ice-station or tented survey team. He loved walking and music. He had taken up the oboe at Winchester and played both oboe and cor anglais in Cambridge for the rest of his life; he began in the orchestras of Cambridge University Musical Society and in the Cambridge Philharmonic, but his greatest love was wind chamber music. With musical friends he played for forty years in a quintet, meeting whenever all were available in the intimate atmosphere of college rooms or home. He lived for almost all those years in Harston, in a beautiful Queen Anne house, with Iris, his wife, and their four children. Many students, colleagues and visitors from the far north will recall with great pleasure the hospitality provided by Terence and Iris at Harston.

Appropriately, his funeral service in the local parish church ended with the singing in Russian of the *Kontakion for the Departed*, and later at a concert in his memory young members of the university included in their programme Mozart's Oboe Quartet in F, K.470.

Bibliography and Sources

1. OBITUARIES OF TERENCE EDWARD ARMSTRONG

Heap, J. (Compiler), *Polar Record*, Vol. no. 32, 182 (1996), 265–70.

Savours, A., (Mrs A. Shirley) *Hakluyt Society*, Annual Report, 1996, 28–30.

Daily Telegraph, 26 February 1996.

Guardian, 29 February 1996.

Independent, 1 March 1996.

The Times, 1 March 1996.

2. SELECTED WORKS BY T.E. ARMSTRONG

1952 *The Northern Sea Route, Soviet Exploitation of the North East Passage*, Scott Polar Research Institute, Special Publication Number 1, Cambridge University Press, Cambridge.

1958 *The Russians in the Arctic, Aspects of Soviet Exploration and Exploitation of the Far North, 1937–1957*, Methuen, London.

1958 *Atlas of Sea Ice North of the USSR*, GB Hydrographic Department, London.

1965 *Russian Settlement in the North*, Cambridge University Press, Cambridge.

1966 (With B.B. Roberts and C.W.M. Swithinbank) *Illustrated Glossary of Snow and Ice Terms*. Scott Polar Research Institute, Special Publication Number 4, Cambridge University Press, Cambridge, (2nd edn, 1973).

1975 *Yermak's Campaign in Siberia* (ed.), The Hakluyt Society, London.

1978 (With G. Rogers and G. Rowley) *The Circumpolar North*, Methuen, London.

3. OTHER WORKS

Numerous articles in *Polar Record*, Scott Polar Research Institute, Cambridge.

4. ARCHIVAL SOURCES

Taped interview by H.G.R. King with Terence Edward Armstrong: an invaluable source of biographical material in the Oral History Collection held at the Scott Polar Research Institute.

PhD Thesis, *The Northern Sea Route*, SPRI.

Peter Speak was for many years Head of Geography at Cambridgeshire College of Arts and Technology (now Anglia Polytechnic University) in Cambridge. Latterly, he has directed the course for the MPhil in Polar Studies at the Scott Polar Research Institute in the University of Cambridge, and is at present Research Associate at that Institute.

Chronology

1920 Born Oxted, Surrey, 7 April 1920, educated at Winchester School

1938 Entered University of Cambridge, Magdalene College, to read modern languages

1940 First class Honours, Part One of Modern Languages Tripos

1940–6	Served in Army Intelligence Corps, commissioned in 1942; served in North Africa, The Netherlands, Germany, Norway
1943	Married Iris Forbes
1947	Returned to Cambridge; appointed Fellow in Russian at the Scott Polar Research Institute
1951	Awarded PhD for thesis on *The Northern Sea Route*
1954	Voyage on HMCS *Labrador* through Northwest Passage
1956	Assistant Director of Research: first visit to Soviet Union
1964	Founder Fellow, Clare Hall, University of Cambridge
1977	*Ad hominem* Reader in Arctic Studies
1982–3	Acting Director of the SPRI
1983	Retired from SPRI; Visiting Professor, Trent University, Ontario, Canada; Chairman of the Natural Environmental Research Council Committee on Arctic Science Policy
1985–7	Vice-President of Clare Hall
1996	Died in Harston, near Cambridge, on 21 February

Charles Franklin Brooks

1891–1958

William A. Koelsch

Reproduced courtesy of the Clark University Archives

1. Education, Life and Work

Charles Franklin Brooks – 'Charlie' to his many friends – was born in St Paul, Minnesota on 2 May 1891, son of Morgan and Frona Marie (Brooks) Brooks. His father, a long-term professor of electrical engineering at the University of Illinois, invented and patented one of the earliest automatic telephone systems. Brooks prepared for college at Thornton High School in Urbana, Illinois, at the Culver Military Academy in Culver, Indiana, and at the Academy of the University of Illinois.

Brooks enrolled as a special student in the University of Illinois engineering programme in 1907, and the following year entered Harvard College as a member of the Class of 1912. Though he completed his undergraduate work early and the AB degree was actually conferred in 1911, when he had just turned twenty, Brooks maintained his life-long Harvard identity 'as of 1912', the year he also obtained the AM degree. From childhood he had been interested in storms, but his decision to prepare himself professionally in meteorology had been stimulated by reading physical geography for the Harvard admissions examinations. William Morris Davis' classic text, *Elementary Meteorology* (1894), taught him methods of observation.

While a graduate student in Harvard's geology and geography programme, Brooks served as a research assistant at Harvard's Blue Hill Meteorological Observatory during the interregnum following the death of its founder, Abbot Lawrence Rotch, in April 1912. This appointment allowed him to read widely in the meteorological literature in Rotch's library and reinforced his broad interests in weather studies. Evidently having run afoul of Mrs Rotch, chief financial supporter of the work pending the settlement of her husband's estate, Brooks resigned his Blue Hill post the following year. He then became a teaching assistant to his mentor,

Robert DeCourcy Ward, and to Wallace W. Atwood, in the introductory physical geography and meteorology courses offered at Harvard and in Radcliffe College. Brooks earned the PhD degree in 1914 with a study of the snowfall of the eastern USA, a field in which he was to have a life-long interest. Brooks took his final examinations just three weeks after his twenty-third birthday. His was only the second doctorate in meteorology and climatology to be awarded by a US university.

No academic position opening for him in 1914, Brooks worked for a year in Washington at the Bureau of Farm Management in the US Department of Agriculture. Following the resignations of geographers Isaiah Bowman and Ellsworth Huntington from Yale's department of geology, Brooks secured a post in that university as an instructor in geography, continuing to work with the Office of Farm Management on a part-time basis. His first full-time teaching work was unsuccessful, however, and his Yale appointment was not renewed when it expired in 1918. Brooks then found temporary employment teaching in the US Army's Signal Corps' wartime School of Meteorology at the Agricultural and Mechanical College of Texas at College Station, Texas. Here he attracted several of his military students into meteorological careers. In November 1918 he joined the US Weather Bureau's headquarters in Washington as a staff meteorologist and Associate Editor of the *Monthly Weather Review*, becoming Editor in July 1919.

Disheartened by post-war cutbacks in bureau funding, Brooks began to look again for an academic position. In 1921, Atwood, now president of Clark University and founder of the Graduate School of Geography there, called him to Worcester as the new school's first full-time faculty member. Atwood had known Brooks since becoming Professor of Physiography at Harvard in 1913, and regarded him as 'a brilliant man' with a promising future. Brooks brought to Clark the fledgling American Meteorological Society, which he had called into being in 1919, as well as its new *Bulletin*, which he edited during its early years.

In September 1931, Brooks was recalled to Harvard by Rotch's cousin, President Abbott Lawrence Lowell, as Professor of Meteorology and Director of the Blue Hill Meteorological Observatory, initially on a half-time basis so that he could finish his autumn courses at Clark. Ward died unexpectedly in November 1931, so that during 1931–2 Brooks taught Ward's Harvard climatology course in addition to his own. Although Brooks regularly supervised the research of advanced Harvard and MIT students at Blue Hill, most of his later career was spent as a scientific administrator. During the long tenure of his predecessor, Alexander McAdie, the scientific standard of the observatory had declined, although McAdie had raised its endowment substantially. Brooks' major achievement at Harvard was to restore and indeed to enhance the observatory's national and international reputation as a centre for meteorological scholarship. In 1934 he launched (and largely edited) a new monograph series, the *Harvard Meteorological Studies*, as an outlet for Blue Hill research.

Lowell's successors, Presidents James Bryant Conant and Nathan M. Pusey, were sympathetic neither to geography nor to atmospheric science. In the same year (1948) that Harvard made the decision to discontinue its geography programme, to which Brooks was attached in his teaching capacity, the university also ended its partial subsidy of Blue Hill's operating costs. By 1950 it appeared also that the Harvard administration was planning to close down the observatory following Brooks' retirement. Brooks and friends of Blue Hill responded by securing a number of research contracts and making significant additions to endowment, thus making the observatory self-sustaining.

After Brooks' retirement, however, the observatory remained open on a limited basis for only a few more years. Since 1959 the National Weather Service has made the daily observations there, giving Blue Hill the longest continuous record (since 1885) on one site in the Western hemisphere. In 1971 Harvard transferred the property, which it had leased since 1896, to the Metropolitan District Commission, and it became part of the surrounding state reservation. The observatory building was placed on the National Register of Historic Sites in 1980 and made a National Historic Landmark in 1989; it is now an earth sciences education centre.

During his academic career, Brooks quickly developed both a national and an international scholarly reputation in climatology, serving on several international scientific commissions and being honoured with election to a number of meteorological and other scientific societies, as well as serving on several National Research Council committees. During the 1920s, Brooks was also a consultant on marine meteorology to the Carnegie Institution of Washington and the Scripps Institution of Oceanography. He was a member of both Phi Beta Kappa and Sigma Xi. He was president of the Association of American Geographers in 1947.

Brooks retired on 31 August 1957. During the night of 7–8 January 1958, snow fell on Blue Hill. The next morning Brooks, whose dissertation research had been on snowfall, died unexpectedly of a heart attack while measuring its most recent occurrence outside his Milton home. By an odd coincidence, his first Harvard PhD, Edmund Schulman, a noted dendroclimatologist at the University of Arizona's Laboratory of Tree-Ring Research, died of a heart attack in Tucson on the same day. Brooks was survived by his wife, Eleanor Stabler Brooks, a frequent collaborator in his work, and by seven children, three of whom had earned doctor's degrees. Both his sons became professors; the elder, Edward M. Brooks, earned an ScD in MIT's meteorology programme and had a distinguished career as a geophysicist.

2. Scientific Ideas and Geographical Thought

Brooks has been called a 'naturalist of the weather', a phrase that indicates his comprehensive interest in observing all manner of weather-related phenomena and also suggests his links with the late nineteenth-century natural history tradition. He might equally, during his earlier years at least, have been called a 'geographer of the weather', in that, like his mentor Ward, he had a primary commitment to the development of a geographical climatology. Brooks had put himself squarely in that tradition in a *Monthly Weather Review* essay in which he argued that, although there was much opportunity for experimental work in meteorological physics, for the most part 'the experiments are conducted by Nature in such a way that man can only observe processes and results', and that therefore the geographer and geologist, who were then largely field-oriented scientists, 'generally know more about the weather than does the physicist' (Brooks, 1918b). In a 1920 letter to Bowman, he refers to himself as a 'backsliding geographer' and complains that he had found meteorology *per se*, at least as practised at that time at the Weather Bureau, to be 'narrow'.

Brooks did not reject the study of meteorological physics, and actively collaborated for several years with the Clark physicist Robert Goddard. But, like Ward's, Brooks' own research often stressed those aspects of the weather that directly affected human activity. Among his many publications were short notes on such topics as 'Phenology: responses of human life to the advance of the seasons', 'Local climates of Worcester, Mass. as a factor in city zoning', and 'Meteorology of an icehouse fire'. One of his responsibilities while at Harvard was to provide the President with a weather forecast for Commencement Week.

Brooks' early research on snowfall was the first major study of its kind. It foreshadowed a life-long commitment to studies in the climatology of New England and that of North America more generally. While at the Office of Farm Management, and on a part-time basis through his Yale years, Brooks worked particularly on the frost maps of the *Atlas of American Agriculture*, and also studied the relation of climate to planting and harvest seasonalities in various parts of the USA. This interest in applied climatology was to be life-long, and was to bring him many opportunities for consulting work.

At Clark, Brooks and his student assistants pursued several lines of research. One of these was a synoptic study of the relation of the sea surface temperatures of the Gulf Stream to the weather of eastern North America and western Europe, with the aim of improving long-range forecasts. This research was supported both by Clark and by extra-mural funds through the American Meteorological Society, and was conducted in cooperation with the Weather Bureau and the Woods Hole Oceanographic Institution. Brooks and his students also made comprehensive studies of New England rainfall and, beginning in 1927, collaborated with Ward on the North American and West Indian portions of the classic Köppen-Geiger *Handbuch der Klimatologie*.

Although Ward had initially drafted the text for the US portion, Brooks and his graduate students were responsible for assembling the initial data from the Weather Bureau and other sources to construct the tables and maps. The climatic maps were subsequently published in folio size and greater detail by Harvard University Press; they remained the best of their kind for many years. Following Ward's death, Brooks became responsible for revising the text and completing all but the Canadian portions of the North America volume. Brooks and his students also conducted studies in New England fire weather, hosting two conferences on this subject at Clark, and made the first detailed studies of ice storms in the region.

After his removal to Blue Hill, much of Brooks' time was spent organizing and encouraging the research of others. His own was eclipsed both by the burdens of institutional administration and fund-raising and by the rise of a newer and more physically based meteorology at the nearby Massachusetts Institute of Technology. His interest in instrumentation and precise measurement, represented in the 1920s by the seawater thermographs used in his marine meteorological studies, continued. In 1932 Brooks organized a network of special weather stations for a study of eclipse meteorology. In the same year he helped secure a permanent high-altitude weather observatory on Mount Washington, in the White Mountains of New Hampshire, serving as president after its formal incorporation and, until 1946, as its meteorological director. Here he supervised studies of the liquid water content and icing rates of clouds. Brooks conceived of the Harvard radiosonde as a device to improve high-altitude observations, and between 1935 and 1942 he sponsored Karl Lange's experiments at Blue Hill with the first routinely used American radiosondes.

During the Second World War, Brooks served as a consultant to the US Army Quartermaster Corps on studies of weather and climate in relation to military needs for clothing and equipment. His long-time expertise in cloud forms as an aid to forecasting bore fruit in the World Meteorological Organization's collaborative *International Cloud Atlas* (1956), a project for which he had furnished the legends and checked and edited the final text. As director of the observatory, Brooks maintained cordial relations with the US Weather Bureau, successfully advocating the establishment of a district forecasting centre at Boston and hosting the Bureau's

Solar Radiation Field Testing Unit at Blue Hill from 1940 onwards. Although himself primarily an observer, Brooks encouraged the newer, more theoretical approaches among his associates as well as advanced instrumental work. Students and faculty from Harvard and MIT found him a willing source of support and encouragement, and the facilities of Blue Hill, including the superb library that Brooks had built up there, a good place for their own research.

3. Influence and Spread of Ideas

Brooks' influence was first of all personal. Himself always open to new ideas, he strongly influenced younger people and touched all those about him with his delight in all aspects of weather studies. This encouragement was reflected in his extensive correspondence with everyone from international colleagues in meteorology to the junior high school student with a science project to be done.

Throughout his mature professional life Brooks was engaged in building popular interest in the weather. For at least its first twenty-five years the American Meteorological Society had a broad general membership and its *Bulletin* could be (and was) read by all interested in weather phenomena. At Clark he established a weather station according to Weather Bureau standards atop the physics building from which students in his course on 'The Passing Weather' made weather forecasts; these were then broadcast over a 300-mile range by the university's high-powered new radio station. When he went to Harvard he established a similar station on the roof of the new Institute of Geographical Exploration building.

Also at Clark, Brooks developed correspondence versions of his courses which were among the most popular offerings of the university's new Home Study Department. One of these students, Jerome Namias, later a leader in American meteorological research, credited these courses with arousing his first serious interest in the subject. Brooks also regularly taught in the Clark Summer School, bringing his subject to many schoolteachers and recruiting some of these for further graduate work in geography and climatology under his direction. When the School of Geography instituted a graduate field course, Brooks inaugurated the tradition (which lasted until the 1960s) of rousting the students out of their warm beds an hour before dawn to make observations of temperature in a field mapping exercise designed to reveal microclimatic variation.

In the summer of 1922 Brooks had encouraged one of his students, John Nelson, a journalist enrolled in his 'Passing Weather' course, to summarize his classroom explanations of the daily weather for the local newspaper. This led to the production of more general notes, prepared with Nelson and later with Eleanor Brooks, which, from May 1923 until April 1927, were circulated to some thirty US and Canadian newspapers by Science Service. Many of these were collected in a popular book under the series title, *Why the Weather?*, subsequently recognized by the American Library Association as one of the most successful contemporary popularizations of science. Beginning in November 1951, Brooks returned to this early popularizing role, broadcasting weekly explanations of 'Why the Weather?' through the new 20,000-watt Boston public radio station WGBH-FM, whose transmitting facilities were initially located on Blue Hill.

Brooks had begun his years of teaching by patterning his courses after those of his mentor, Ward, even to proposing an anthropoclimatology course on 'Climatic environments of the white race' at Clark. But, unlike Ward, Brooks moved with the times and pushed his own graduate students towards a knowledge of physical climatology. Beginning in 1925, Brooks offered advanced courses in laboratory meteorology and in the physics of the air (with stiff scientific and mathematical prerequisites) jointly with Clark physicist Robert Goddard, who also offered a graduate research course in meteorology. Brooks subsequently suggested Roswell, New Mexico, as the optimum site climatically for Goddard's experiments with high-altitude, liquid-fuel rockets, after the physicist had been barred by the Massachusetts fire marshal from conducting any further experiments within the state.

Brooks' courses in meteorology and climatology attracted both graduate and undergraduate students. When one student at another college decided he wanted to become a professional weather forecaster, a Weather Bureau official recommended Clark as 'the best college in the East' for his career preparation. Other Clark students and alumni joined him later as assistants at Blue Hill. His varied research projects gave his students and associates hands-on experience, as well as topics for their own apprentice research and occasionally for their first publications.

Of the 37 geography doctorates completed at Clark during his 11 years on the faculty, Brooks directed five on meteorological or climatological topics. Two others who studied with Brooks later took doctorates at Clark in climatology, and acknowledged Brooks' continuing help on their dissertations. Brooks' courses also prepared graduate students in regional geography for the climatological portions of their theses and dissertations. In addition, Brooks supervised several MA theses.

Brooks went to Harvard in 1931 with the hope that the early lead taken by the Harvard department of geology and geography in the days of Davis and Ward in the fields of meteorology and climatology could be renewed. Meteorology proper would, he argued, be taught in the new programme at MIT, but Harvard might appropriately offer a cultural course in meteorology and climatology, graduate courses in climatology, and research opportunities in both areas at Blue Hill. No one was appointed to take Ward's place as Professor of Climatology, however, though his climatology courses continued to be listed in the Harvard catalogue for some years after 1931. Brooks' professorship in meteorology had been, since its establishment in 1906, primarily a courtesy title held by the director of Blue Hill, and its previous occupants had offered only specialized training. In any case, during the early 1930s Brooks was preoccupied with the administrative tasks of rehabilitating Blue Hill and with other projects.

Brooks and other Blue Hill staff members were active participants in the joint MIT–Harvard seminar in meteorology held weekly on the MIT campus. The graduate students in Brooks' reading and research courses were required to attend these seminars, ensuring that the Harvard geographers and geologists would be exposed to the newer Scandinavian methods and approaches being expounded by Carl-Gustaf Rossby and his associates at the latter institution. A refugee scholar, Victor Conrad, formerly Professor of Meteorology and Climatology at the University of Vienna, served as the Robert DeCourcy Ward Research Associate in Climatology from 1939 to 1951 and collaborated with Brooks in supervising graduate research and by giving occasional instruction in physical climatology.

In 1939–40, Brooks was able to reconstitute a foundation programme in climatology by getting Hurd Willett of the MIT faculty to teach a one-semester general meteorology course each autumn to Harvard undergraduates. This was to be followed in alternate years by a semester course in regional climatology offered by Harvard geographer Edward A. Ackerman, who had earned an MA in climatology under Brooks, and a course on the climatology of North America by Brooks himself. This programme was soon to be disrupted, however, by Ackerman's frequent absences on wartime service in Washington and by the diversion of Brooks and other Blue Hill staff members to war-related meteorology courses. His first PhD, Edmund Schulman, received his degree only in 1944, the first Harvard doctorate in climatology since 1917.

In the post-war period, Brooks and the Blue Hill Visiting Committee argued strongly for expansion of Harvard's climatology teaching programme. But given the shrinkage of the geography programme and his own preoccupation with Blue Hill's financial and administrative problems, Brooks (with, for a time, Conrad) was able to offer instruction in climatology only in tutorial and directed research courses, taken by only one or two Harvard undergraduate or graduate students a year. Harvard's division of geology, during the 1950s, was unwilling to make the necessary financial provision for even a minimum set of cultural courses in meteorology and climatology on the main campus.

In the period between 1944 and his retirement in 1957, Brooks directed only two more dissertations; his last student received the doctor's degree at the commencement following Brooks' death. Brooks' careful training of the many young and short-term observers and assistants at Blue Hill and at the Harvard weather station, however, was a form of teaching in which his dynamic personal influence was deeply felt by a wide range of students, some of whom became professional meteorologists. Brooks also advised research students at other institutions; in 1949, for instance, he reported that during the previous year he had collaborated on eight theses at other universities, four of them at Clark.

From the standpoint of producing a regular stream of geographical climatologists who could revive or maintain a robust tradition within geography, however, Brooks' move from Clark to Harvard had proven unsatisfactory for both institutions. Under Brooks' successor, Samuel Van Valkenburg, courses in this area at Clark dwindled to four and then three, mostly descriptive treatments of world regional climatology. At Harvard, Willett found that, as he increased the mathematical and physical expectations of his introductory meteorology course to meet the needs of physicists and engineers, the number of geography students declined.

When Brooks retired in 1957, he was the last surviving geographer attached to the Harvard geology and geography department, and his departure effectively ended geographical climatology at Harvard. The only other tenured geographer surviving President Conant's 1948 decision to discontinue the geography programme, the human geographer Derwent Whittlesey, had died the previous November. In the interim, the two geologists most closely associated with geography, the physiographers Kirk Bryan and Kirtley Mather, had retired, in 1950 and 1954 respectively; Harvard's Institute of Geographical Exploration had been discontinued in 1951; and Blue Hill had lost its operating subsidy. Shortly after Brooks' retirement, a process of dismantling Blue Hill operations began. The library and meteorological data collections Brooks had nurtured were broken up, and the Blue Hill endowment was diverted to the support of a Harvard faculty member. The research of Brooks' successor was largely on the physics of the upper air within Harvard's Center for Earth and Planetary Physics, and links with applied physics and oceanography replaced the older ties with geology and geography.

Brooks was remembered and loved by his associates and students for his kindness, his goodwill and open-mindedness, his moral and sometimes his financial support for them and their work, and his patience in the face of both human and institutional limitations. Brooks' most lasting legacy, however, is the American Meteorological Society (AMS), of which he was the major founder, and which he had nurtured as secretary and editor through the days of small things. At the time of his death the AMS had become the largest and most active such society in the world. Although the AMS has largely lost the popular character that Brooks initially envisioned for it, he is still commemorated there by the Charles Franklin Brooks award for distinguished service to the AMS and in the Charles F. Brooks Memorial Library at the AMS's elegant Bulfinch-designed mansion overlooking Boston Common.

Bibliography and Sources

1. SELECTED REFERENCES ABOUT CHARLES F. BROOKS

'Brooks, Charles Franklin', *National Cyclopedia of American Biography*, Current Volume H (1952), 332–3.

Conover, J.H., *The Blue Hill Meteorological Observatory: The First Hundred Years, 1885–1985*, American Meteorological Society, Boston, 1990. (Contains bibliography of Blue Hill research.)

Conover, J.H. 'Charles Franklin Brooks, Meteorologist', *Science*, Vol. 127 (1958), 863–4.

Conover, J.H., 'Dr. C.F. Brooks', *Quarterly Journal of the Royal Meteorological Society*, Vol. 84 (1958), 484.

Houghton, H.G., 'Charles Franklin Brooks, 1891–1958', *Bulletin of the American Meteorological Society*, Vol. 39 (1958), 155.

Houghton, H.G., 'Prof. Charles F. Brooks', *Nature*, Vol. 181 (1958), 1440.

Koelsch, W.A., 'From geo- to physical science: meteorology and the American university, 1919–1945', in Fleming, J.R. (ed.), *Historical Essays on Meteorology, 1919–1995: The Diamond Anniversary Volume of the American Meteorological Society*, American Meteorological Society, Boston, 1996, 511–40.

Mather, K.F., Conover, J.H. and Miller, J.P., 'Charles Franklin Brooks', *Harvard University Gazette*, Vol. 53, no. 37 (1958), 191–3.

Stone, R.G., 'Charles F. Brooks', *Geographical Review*, Vol. 48 (1958), 443–4.

Van Valkenburg, S., 'Charles F. Brooks, 1891–1958', *Annals of the Association of American Geographers*, Vol. 49 (1959), 461–5 (includes selected bibliography).

Manuscript material may be found in the archives of the institutions with which Brooks was connected, especially: US National Archives, Record Group 27, Records of the Weather Bureau; Clark University Archives (Wallace W. Atwood Papers); and Harvard University Archives (principally Records of the Blue Hill Meteorological Observatory, 1884–c.1960, especially Correspondence of Charles F. Brooks, Director, 1931–1957). A small but important body of material is at the American Meteorological Society. Brooks' world-wide network of correspondents means that letters will appear in many related manuscript collections, e.g. the archives of the American Geographical Society, the Harry Wexler Papers in the US Library of Congress, and the Ellsworth Huntington Papers in the Yale University Library, to name but three. A portrait by A. Gourine, painted at the time of Brooks' retirement, hangs in the Gordon McKay Library, Pierce Hall, Harvard University; a copy by the same artist is in the Brooks Library of the American Meteorological Society.

2. SELECTED BIBLIOGRAPHY OF WORKS BY CHARLES F. BROOKS

1913 'Snowfall of the United States', *Quarterly Journal of the Royal Meteorological Society*, Vol. 39, 81–6 (part of MA thesis).

1915 'The snowfall of the eastern United States', *Monthly Weather Review*, Vol. 43, 2–11 (part of PhD thesis).

1917a 'Island Nantucket', *Geographical Review*, Vol. 4, 197–207.

1917b 'New England snowfall', *Geographical Review*, Vol. 3, 222–40; also *Monthly Weather Review*, Vol. 45, 217–85.

1918a 'The "Old Fashioned" winter of 1917–1918', *Geographical Review*, Vol. 5, 405–14.

1918b 'Collegiate instruction in meteorology', *Monthly Weather Review*, Vol. 46, 554–60.

1919 'General extent of collegiate instruction in meteorology and climatology in the United States', *Monthly Weather Review*, Vol. 47, 169–70.

1924 (with J. Nelson and E. Stabler Brooks) *Why the Weather?* Harcourt, Brace & Co., New York; revised and enlarged edition 1935.

1930a 'Gulf Stream studies: general meteorological project', *Monthly Weather Review*, Vol. 58, 103–6.

1930b (with J. Henry Weber and G. Richards) 'The rainfall of New England', *Journal of the New England Water Works Association*, Vol. 44, 1–118 (reprinted as separate volume).

1933 'How may one define and study local climates?', *Comptes Rendus du Congrès International de Géographie* (13th), Paris, 1931, Tome 2 fasc. 1, Armand Colin, Paris, 291–300.

1934 (with R.DeC. Ward) 'Climatology of the West Indies', in Köppen, W. and Geiger, R. (eds), *Handbuch der Klimatologie*, Vol. 2, Part I, Verlag von Gebrüder Borntraeger, Berlin.

1936a (with A.J. Connor *et al.*) Climatic maps of North America, folio of 26 maps published by Harvard University Press for the Blue Hill Meteorological Observatory.

1936b (with R.DeC. Ward and A.J. Connor) 'The climates of North America', in Köppen, W. and Geiger, R. (eds), *Handbuch der Klimatologie*, Vol. 2, Part I, Verlag von Gebrüder Borntraeger, Berlin.

1937 (with A.H. Thiessen), 'The meteorological causes of great floods in the United States', *Geographical Review*, Vol. 27, 269–90; reprinted, Smithsonian Institution, *Annual Report for 1938*, 325–48.

1939 'Hurricanes into New England: meteorology of the storm of September 21, 1938', *Geographical Review*, Vol. 29, 119–27; reprinted (with revisions), Smithsonian Institution, *Annual Report for 1939*, 241–51.

1940 'The worst weather in the world', *Appalachia*, New Series Vol. 6, 194–202.

1942 'Some North American connections of Caribbean climate', *Proceedings of the 8th American Science Congress*, Vol. 7, *Physical and Chemical Sciences*, 297–311.

1943 (with W.B. Liverance, Jr) 'Cloudiness and sunshine in New England', *Bulletin of the American Meteorological Society*, Vol. 24, 263–74.

1947 (with E.M. Brooks) 'Sunshine recorders: a comparative study of the burning glass and thermometric systems', *Journal of Meteorology*, Vol. 4, 105–15.

1948 'The climatic record: its content, limitations and geographic value', *Annals, Association of American Geographers*, Vol. 38, 153–68 (presidential address).

1951 'The use of clouds in forecasting', in Malone, T.F. (ed.), *Compendium of Meteorology*, American Meteorological Society, Boston, 1167–78.

1956 'The Mount Washington Observatory', *Weatherwise*, Vol. 9, No. 5, 151–4.

1958 (with S.J. Richardson) 'The Blue Hill Meteorological Observatory Library', *Harvard Library Bulletin*, Vol. 12, 271–81.

William A. Koelsch is Professor of History and Geography, Graduate School of Geography, Clark University, Worcester, MA 01610, USA.

Chronology

1891	Born on 2 May in St Paul, Minnesota
	Student at Thornton High School, Urbana, Illinois
1905–6	Student at Culver Military Academy, Culver, Indiana
1906–7	Student at the Academy of the University of Illinois, Urbana
1907–8	Special student, University of Illinois
1908–11	Student, Harvard College, Cambridge, Massachusetts
1911–14	Graduate student, Department of Geology and Geography, Harvard University
1912–13	Research assistant, Blue Hill Meteorological Observatory
1913–14	Assistant in meteorology and physiography, Harvard College; assistant in physical geography, Radcliffe College (Spring 1914)
1914	Married on 4 June to Eleanor Merritt Stabler
1914–15	Assistant, Office of Farm Management, US Department of Agriculture, Washington, DC.
1915–18	Instructor in geography, Yale University, New Haven, Connecticut
1918	Assistant instructor in meteorology, US Army Signal Corps School of Meteorology, College Station, Texas (18 May to 7 November)
1918–21	Meteorologist, associate editor and editor, *Monthly Weather Review*, US Weather Bureau, Washington, DC.
1919	American Meteorological Society (secretary, 1919–54; editor, *Bulletin*, 1920–7, 1929–36; honorary secretary, 1954–8)

1920	Vice-president, Association of American Geographers; elected Life Fellow, Royal Meteorological Society
1921–32	Associate professor and (from 1926) professor of geography, Graduate School of Geography, Clark University, Worcester, Massachusetts (to February 1932)
1924	Publication of *Why the Weather?*
1931–45	Member, International Climatological Commission
1931–57	Professor of meteorology, Harvard University, and director, Blue Hill Meteorological Observatory
1932	Co-organizer, Mount Washington Observatory (Meteorological director, 1932–46; president, 1936–58)
1934–6	Publication of sections of Köppen-Geiger *Handbuch der Klimatologie*
1935–8	Chair, Section on Meteorology, American Geophysical Union
1936–48	Member, International Commission on Snow and Glaciers
1939	Visiting professor, The University of Chicago (Summer Quarter)
1947	President, Association of American Geographers
1947–53	International Commission for Instruments and Methods of Observation; International Commission on Clouds and Hygrometers
1956	Publication of International Cloud Atlas
1957	Professor of meteorology, emeritus (from 1 September)
1958	Died on 8 January at Milton, Massachusetts

Shiba P. Chatterjee

1903–1989

Sitanshu Mookerjee

If there was one Grand Mughal who ruled over the domain of geography in India for nearly four decades, his name was undoubtedly S.P. Chatterjee. The cause that he held dearest in life was that of the promotion of geography, both in academia and in nation-building. Not only was he the undisputed doyen of Indian geographers, but his reputation travelled far and wide: eventually he became the first Asian to be elected president of the International Geographical Union.

In this role of IGU president, he established an unsurpassed record of having visited practically all adherent countries of the IGU, including South Africa, which was at that time 'taboo' for an Indian citizen, because of the apartheid laws. He also organized the 21st International Geographical Congress in 1968, along with the 11th General Assembly of the IGU, and the 3rd General Assembly of the International Cartographic Association, in New Delhi. This was the first time the congress had been held in Asia. The congress was held in the winter month of December in deference to the fury of the Indian monsoon, and enlarged to include a number of pre- and post-congress symposia, which included a range of exciting excursions to different cultural sites. A number of regional monographs were followed by several volumes of proceedings and papers, all of which bore his stamp. The presidential address that Professor Chatterjee delivered at this 1968 congress (in the presence of Prime Minister Indira Gandhi) bore the title 'Towards global peace and harmony', and attempted a rapprochement between developing and developed countries. On its publication, it was acclaimed as being a remarkable piece of geographical literature.

Professor Chatterjee was not only the recipient of several prestigious international awards, but was also the only Indian geographer to be decorated by the president of India. In 1985 he received one of India's highest Republic Day awards – 'Padma Bhushan'. He thus brought geography to the attention of Indian society. His was an immensely colourful career: the key to his success was his total dedication to a cause.

Yet in his private life, Chatterjee was unaffected and soft-spoken, gentle and genial, humane and hospitable.

1. Education, Life and Work

Shibaprasad Chattopadhadhya (his full name in mother-tongue Bengali) was born in Calcutta, India, on 22 February 1903. He was educated at Banares Hindu University at Varanasi, on the Ganga (Ganges) in Uttar Pradesh, from which he obtained his MSc in geology in 1926. Thereafter he had four years as head of the Department of Geology and Geography at the University of Rangoon, Burma (then part of India), from 1928 to 1932. It was during the tenure of this post at Rangoon that Chatterjee came to know the celebrated British geographer, L. Dudley Stamp (later Professor Sir Lawrence), whose love of the Burmese countryside prompted him to dedicate his monumental volume on Asia to his wife: 'In memory of bullock-cart days and Irrawaddy nights.'

After these four years of valuable teaching experience, Chatterjee proceeded to Europe for further studies. He went first to Paris, where he earned his doctorate in geography from the Sorbonne, working under the famous French geographer, professor Emanuel de Martonne. His thesis, *Le Plateau de Meghalaya*, was published in Paris in 1936. Having completed his research in France, he immediately journeyed to England in pursuit of his other interest, pedagogy. He received a teacher's diploma from the University of London, and then proceeded to a PhD in education at the same university, basing his work on a comparative study of the educational systems of England and France.

Chatterjee returned to India in 1937 to take up a teaching position with the Faculty of Education at the University of Calcutta, and with great enthusiasm organized a teachers' training course in geography. This course attracted schoolteachers from the whole of Bengal Province and exposed them to modern geography, which soon became a compulsory subject in high schools. Simultaneously, he was responsible for a geography department in Calcutta University. Many of his students on the teachers' training course had already shown a keen interest in geography, and consequently he introduced the Honours and postgraduate courses (MA/MSc) into Calcutta University from 1939, as lecturer-in-charge of the newly established department. Although the Aligarh Muslim University already had a programme of postgraduate teaching in geography that had produced several batches of students, its impact at that time was rather limited. It was the establishment of the Calcutta department that brought geography to 'take-off stage' in India.

In those days (and even into the early 1940s), Chatterjee often lamented that geography was often treated as a Cinderella among university subjects. Perhaps this was the picture at an early phase of the development of geography as an academic subject in a number of countries. But Cinderella eventually triumphed when she met her Prince Charming! In the case of Indian geography, the hapless lass's Prince Charming was the tall and handsome Chatterjee. She has not looked back since. After some initial teething problems, geography has seen phenomenal growth, not only as a rapidly expanding university discipline, but through its involvement in planning-related enterprises – at national, regional and urban levels.

Another invitation came to Chatterjee in 1957, from his first love, Rangoon University, to organize a separate postgraduate department of geography as a professor – a rank still denied to him in Calcutta. Chatterjee's belief was that the only alternative to hard work was harder work, and he responded to the Rangoon call. The sojourn was, however, a short one, only until 1959, when Chatterjee returned to Calcutta University, which by then had created a full professorship – justifiably, but rather belatedly, offered to him. He then held the chair until his retirement, after which he was honoured as a Professor Emeritus. During his long and rewarding career at Calcutta University, he served as Dean of the Faculty of Science for two terms, and also as a member of the University Senate and the Syndicate.

The motto of the University of Calcutta is 'Advancement of Learning', but advancement was not always a smooth process. The old guard in the 'hard-core' sciences often looked on geography as something of an interloper. But despite apathy from some quarters, Chatterjee was able to bring geography under faculty guidance, and he developed excellent laboratories and research facilities, equipped with a variety of drawing and surveying instruments, a huge collection of topographic, geological and weather maps of various scales, wall maps and atlases, besides a well-stocked seminar library. These were initially scattered around the main College Street campus of Calcutta University, but later, in the early 1960s, when the University College of Science got another seven-story building constructed on its Ballygrange campus, geography was allotted a whole floor which could accommodate all these laboratories, research facilities, the library and office and lecture halls of varying sizes, and even allow room for the Geographical Society of India's office and library. (The society had been founded as the Calcutta Geographical Society in 1933.) This became perhaps the most spacious and best-equipped geography department in India. Chatterjee, of course, had to work hard, and to fight for every unit of this infrastructure. In the space of just a few years the department developed a broadly based teaching programme, with specialized Honours and Masters' courses in climatology, oceanography and biogeography, as well as economic and political geography.

As the result of Chatterjee's pioneering work in the compilation of the *Atlas of Bengal* (see below), Prime Minister Nehru, in a letter dated May 1953, suggested that Chatterjee should embark on a similar publication for the whole of India. Greatly encouraged by this, Chatterjee began formulating a proposal for the compilation of a *National Atlas of India*. Nehru was impressed, and endorsed the scheme for appraisal and a follow-up by the Planning Commission. All these formalities took some time, and a pilot office was set up in the Geography Department of Calcutta University in about 1956. The National Atlas Unit in the department was eventually upgraded to the fully fledged National Atlas Organization, with government of India support and funding for personnel and equipment. Staff numbers expanded, and by about January 1957 Chatterjee had been appointed Honorary Director, a position he held in addition to his full professorship at the university. In the record time of less than a year from data collection and drafting to the final printing of maps, the preliminary Hindi edition (with English translation in appropriate places) was published by October 1957. It was acclaimed by Indian planners and politicians, and by the community of geographers abroad. In recognition of this work, the Royal Geographical Society of London presented him with the coveted Murchison Award in 1959. Then began the work for the main (English language) edition of the *National Atlas of India*, with its 300 maps: this continued for years at a rather different pace, and the formidable eight-volume atlas finally appeared in 1982. This work was interspersed with the production of a number of other atlases, commissioned by various government ministries, such as Irrigation (1970), Tourism (1976) and Forests (1976). After his retirement from Calcutta University in 1967, Chatterjee continued as Honorary

Director of the National Atlas Organization for some years, now devoting his full-time energy to its work. In about 1970, Chatterjee's connection with the organization terminated officially, but he remained active, continuing to guide activities in his slightly altered role of chairman of its advisory committee.

During this period, in the 1970s and beyond, Chatterjee was not bound by any tight schedules in Calcutta, or indeed in India, and it was possible for him to accept short or long teaching assignments abroad. He did stints as visiting professor in widely scattered locations, including Athens (Georgia, USA), Texas, Paris, Heidelberg and Moscow. In the UK he delivered lectures in Liverpool, Nottingham, Lancaster and London. Finally, he held a longer-term visiting professorship at California State University (Los Angeles), which lasted until 1972. (Interestingly, leaving behind the pleasant Californian winter, he was to travel to snowbound Pittsburgh, Pennsylvania, over Christmas, to play bridge at the residence of his old National Atlas colleague Lakshmanan and his Yugoslav wife.)

On his final retirement from US universities he returned to Calcutta, where in the 1960s he had built a spacious house, not too far from the university campus. Here he set up not only his personal library, but also a cartographic office where his assistants came to work for him, for he was at that time contracted to publishing houses to write textbooks. Work was worship for Chatterjee, and he was the last man to decline work. He produced quite a number of texts for higher secondary and junior college use, besides contributing articles to national and international journals. He continued to deliver inaugural or keynote lectures to high-level seminars and conferences all over the country. In fact, all through the 1970s, the present writer, who used to work in the Indian mid-west, 1000 km away, had many occasions to be associated with Chatterjee on committees and at conferences: these included the Conventions on Scientific Terminology at Poona, Jamalpur and Delhi; the All-India Scholars Conference in Bombay (1974); the Bangladesh Geographical Congress at Dhaka (1976); and the National Symposium on the History of Geographical Thought at Najpur (1977). At each of these, the present author was able to discuss geographical issues with Chatterjee. At his house he entertained foreign geographers who occasionally dropped in at Chatterjee's residence when visiting India. Despite his frailty – as a result of surgery – he came down the stairs to open the door and again to see them off on departure. He was always hospitable.

His wife Gouri, who provided great encouragement to Chatterjee in his struggling years, passed away in 1963. His four daughters got married and lived elsewhere – one in Britain, and one as far away as Brazil. In his final years he thus lived alone, but full of ideas. He died on 27 February 1989 at the age of 86 years.

Those who worked with or under him for long or short periods would testify to his unsurpassed capacity for uninterrupted work – no matter what festivities might be taking place around the corner! He set himself a punishing schedule and expected others to do the same. He set his goals high, and would seldom compromise. He was a perfectionist, a stickler for detail, and he used to run through proof after proof before an entry was finally published: many times the printing press was telephoned, just as a 'final proof' was on its way, with the message that yet another proof would shortly appear!

Another characteristic was Chatterjee's fascination with, even craving for, modern technology. He wanted to modernize his drawing office, with the latest equipment from Europe. He was enchanted with the prospects of automation, in the 1950s, long before computers were available, and tried to set up and maintain a state-of-the-art cartography laboratory, not always with instant or complete success. In this high-cost and high-technology endeavour, the present writer spent months in correspondence with German firms in order to acquire modern cartographic equipment.

Chatterjee, the robust scientist who could work long hours, day after day, and could make others follow his example, was also a deeply religious man, well-versed in the 'Shastras' and the scriptures of the Hindus. He was a product of three 'gharanas' or traditions or schools of learning: the 'Santipur' gharana of his early upbringing, the 'Varanasi' gharana of his maturity in the Brahminic culture, and the French gharana, the source of his abiding interest in humanity in its environment and 'Géographie Humaine'. He might have been initiated in the Vaishava cult at his ancestral home at Santipur, Bengal, but his longer stay at Varanasi must have deepened his thirst for the Sanskrit classics. Later in Europe he was greatly influenced by the French concept of the region, as carved out by nature, and modified by humanity. This regional concept had so much entered into his mind that while working on the Garo-Khasi-Jaintia Hills of Assam for his doctoral thesis in France, he visualized there a perfect '*pays*' to which he gave the sonorous Sanskrit name 'Meghalaya' ('abode of the clouds' – where else could it be than in the vicinity of Cherrapunji, the world's wettest place?).

His early years and youth spent on the banks of the Sacred Ganga, both at Santipur and Varanasi, had built up the basis of his apparent religiosity. Every morning, say some intimates, he would recite from the holy texts of the Gita or Chandi, or from Markandeya Purana, which he claimed would grant him the mental power he needed in all his troubles. He also performed customary rituals at home.

In his daily life he almost always appeared a tower of strength, never baffled by anything that stood in his way. His philosophy of life was enshrined in the Gita, which commanded man to do his duty steadfastly, and never for a moment to look for the fruits of his labour. 'You have a right to work only, but not to its outcome,' it exhorted further. 'If you can imbibe this philosophy in your working life, you will have attained eternal bliss,' said the Gita. 'If rewards come your way, well and good, but if they don't, never should you worry, but only carry on your task at hand.' This is what Karmayoga is all about, and Chatterjee was deeply immersed in this thought. Devotion and dedication were two key words in his philosophy of life.

One hesitates to ask the question, but was he able to stay on top at all times, without a slip anywhere? Of course, as a human being he did occasionally suffer the odd pitfall. In his desperate pursuit of his goal – the elevation in status of the hitherto neglected subject of geography – he used to encounter many a hindrance from those jealous of him. At times he faltered and chose the 'softer option' for an immediate benefit, or cut short an ongoing debate. In such conflicts between ends and means, Chatterjee occasionally slipped, missing the wood for the trees – to an extent that once emboldened a close colleague and ex-student to remind his professor that one's national character was more important than a national atlas. Occasionally, he showed, in unguarded moments, a propensity to subdue a rival, suppress a fact or even become ruthless with an unsuspecting colleague. These episodes would be known only to those who worked in close proximity to him – to those out in the gallery, Chatterjee was invariably a demi-god. The discordant remarks outlined here are not in any way intended to bring down Chatterjee from his high pedestal, but only to point out that, after all, he was a man, with human frailties.

For his various achievements, Chatterjee received honours and awards aplenty. Apart from the Murchison Award of the Royal Geographical Society in 1959, he had, in his student days, been awarded the Gaudy Medal of the Société de Géographie Commerciale de Paris in 1937. He was also decorated with the Nehru Medal of the National Geographic Society of India in 1963. In recognition of his life-long contributions to geography, several academic and scientific bodies offered honorary memberships or fellowships to him. These included the American Geographical Society of New York, the Chinese Association for the Advancement of Science (Taiwan) and the Association of Japanese Geographers, in 1980. The climax came, however, in 1985, when the government of India bestowed upon Chatterjee one of the country's highest civilian awards – 'Padma Bhushan' – with which he was decorated by the president of India on Republic Day 1985, poignantly just a few years before his death. In failing health, and in a frail physical state, the former tireless world traveller made a trip to New Delhi, after many years of remaining indoors, to receive this award in person.

2. Scientific Ideas and Geographical Thought

Though he was a teacher at heart, Chatterjee's mind was never at rest, for he believed in a positive role for geography, not only in the universities, but also in national development. Geography, he felt, should be an active discipline, and should set out to do something for the good of humanity. In short, application was the *summum bonum* of his geographical philosophy. He wanted to usher in an era in which the geographical profession could demonstrate its skill and capacity to foster the well-being of the country. In national planning, geography could and should play its legitimate role. This became the thrust of his presidential address at the Geography Section of the Madras Session of the Indian Science Congress held in 1941. He then argued that for the proper delineation of planning regions (a task eventually passed to Chatterjee by the Union government much later) and the utilization of all resources therein, what was needed most was a series of maps, covering the entire country, drawn on a uniform, large scale for comparison. By the eve of the partition of the sub-continent he had already prepared a map of Bengal Province showing the distribution of population based on religious affiliation. This map attracted the notice of political parties as the Indian sub-continent stood at the crossroads of partition. Chatterjee also published an 81-page monograph on the partition of Bengal in 1947.

Soon afterwards, from 1950 onwards, he carried out a Land Use Survey of the Howrah District of West Bengal using similar techniques to those of the British Land Utilization Survey conducted by his friend and former colleague Professor L. Dudley Stamp in the 1930s. His resources were limited, but his enthusiasm almost limitless: he was able to inspire a band of schoolteachers, who teamed up to help with the survey during vacations.

Chatterjee was never a man to be inactive at a time of national crisis: around the time of independence and partition he was already working on a series of maps showing the physical basis, as well as the economic resources, of Bengal province. A few faithful friends, such as the late Amit Dutt, who did the tiniest lettering at the cost of his eyesight, assisted in the task of getting the maps drawn. Thus the unique volume *Bengal in Maps* was ready for publication by Orient Longmans just after the province was politically partitioned (into West Bengal and East Pakistan). Chatterjee was able to insert an Appendix showing the new international boundary, conforming to the Ratcliffe Award, with every village abutting onto it. This was a breakthrough, a real watershed, for the geographical profession in India, and one that attracted wide attention, including that of India's first and forward-looking prime minister, Pandit Nehru (1889–1964), and the National Atlas Organization (NAO) was born (see above), this becoming the National Atlas and Thematic Mapping Organization (NATMO) in 1978. NATMO, now linked with the Remote Sensing Agency in Hyderabad, and able to harness satellite imagery and the latest multi-colour printing methods, now employs over 500 persons, including about 350 geography graduates, and is one of the world's leading mapping organizations.

Also at the national level, he was invited by the government to delimit the planning regions of India, and to contribute the chapter on physiography for the new *Gazetteer of India*, both in the 1960s. He was also invited to become a member of the Technical Committee for the Diagnostic Survey of the Damodar Valley Region, for whom he compiled an excellent planning atlas, of 31 plates, in 1969. He was later invited to contribute chapters to a number of prestigious publications in India and abroad, including the entry on the Himalayan mountain range for the *Encyclopedia Britannica* (Chicago, 1974), and the chapter on India in the *World Atlas of Agriculture*, published in 1973 by the International Association for Agricultural Economics. He produced the volumes on *Progress in Geography* for the Indian Science Congress Association (Calcutta, 1964–8 and 1973), and the valuable report *Geography in Indian Universities* was prepared under Chatterjee's chairmanship by the University Grants Committee (New Delhi, 1968). In his retirement he wrote college textbooks.

His life-long companions in his geographical teaching and research were cartography and thematic mapping. It was Chatterjee who brought geographical cartography and thematic cartography into the limelight in India, and it was as the result of his efforts that the status of cartography was raised. His method of research in this field consisted of the intensive collection of primary data, and its topical classification and quantitative analysis, and the drafting of maps using appropriate techniques and on a suitable projection, each map to be finally summed up in a lucid narrative account. Those of his colleagues, the present writer included, who worked with him during the compilation of India's first national atlas were completely overtaken by the speed at which he could write up the most readable explanatory notes for each of the plates published.

Another of his guiding principles is seen in the way in which he always wanted to work with village-level data – not easy to obtain for the hundreds of thousands of Indian villages. Available official statistics at the district level he would accept only with some reservation, as he believed that one could only generalize from the widest possible database. This meant additional hours of work, and additional visits to libraries and offices: he was dissatisfied if he missed just one of the building-blocks that made up the country's mosaic.

3. Influence and Spread of Ideas

In his long career as a teacher and professor of geography, his primary interest was in geomorphology, having been trained in the best traditions of French landscape studies. He was at his best delving into the topographic complexity of the Indian sub-continent – from the Western Himalaya to the Bay of Bengal, but especially the Meghalaya Plateau. He was also concerned with changes in sea level and coastal morphology. His basic postgraduate degree in geology provided a strong background for the study of landforms. He was equally proficient in variations in lithology, and the ramifications of the Davisian erosion cycle. In his postgraduate classes at Calcutta University, therefore, he proved himself an excellent teacher, especially of geomorphology; many students found the grounding he provided in physiography advantageous when proceeding to higher studies overseas.

But as a pioneer teacher in a newborn department, Chatterjee had to teach other courses too, particularly economic geography, the geography of minerals providing a clear link. As a modern geographer, concerned with current political affairs, he published some critical monographs, e.g. on the frontiers of Poland, and Turkey and its problems.

In his work on the *National Atlas of India*, Chatterjee had developed an expertise and generated enthusiasm almost unparalleled in the history of cartography. He believed that geography's primary application came through maps, so useful for planning in any developing country. Few countries, developed or undeveloped, with the possible exceptions of France and the Soviet Union, had previously conceived or published national atlases on the scale that India did. The idea sparked, and spread through the work of the IGU Commission on National Atlases (of which Chatterjee was an active member). Many more countries followed the lead, and a whole series of national atlases came out.

Through his influence, postgraduate diploma courses in geographical cartography were established in several Indian universities. He had a powerful effect on all those with whom he came into contact, and his indirect effect on Indian geography was immense. However, there were some of his students, research assistants and colleagues who were uncomfortable with his style of working. Chatterjee had a dream, and he was often in a hurry to reach his goal. Not everybody could rise to the occasion, or adhere to his pattern or procedures, and sometimes enthusiasm would flag. Chatterjee found this difficult to understand; it seemed that there was some kind of inner compulsion always driving his spirit. At the end of the first national atlas fever, or soon after, some members of the all-cohesive, all-submissive, all-sacrificing young team, drawn from universities all over India, began to drift away. Some were in search of pastures new, some in pursuit of higher studies, some looking for a more comfortable (or rather less hazardous) career. Lakshmanan moved to Ohio, Mishra to Maryland, Sirha and Bose to Karhataka, Taher and Manotosh to Assam, Mazumber and Das to Delhi, Sulakha and Savitor to women's colleges, and the present writer back to teaching in Illinois. Thus did Chatterjee's influence spread. But some of the old faithfuls, stalwarts like S.P. Dafrupta, G.K. Dutt and A.K. Kunden, stuck to their posts and rose to be directors themselves in their turn.

In his later life, Chatterjee admitted, in a private conversation with this writer, that he had once believed that one could produce substantial, quality work only if one were given adequate funds and staff. But of late he had come to think quite otherwise: it was the devotion of a small group, even with limited resources, that could deliver the goods! Chatterjee expected the world from his students, and on the whole they did not fail him. Indeed, he was fortunate in having produced a golden crop of distinguished students who subsequently built up the first departments of geography in several universities in India – Banares (Varanasi), Nagpur, Osmania (Hyderabad), Karhataka (Dharmabad), Gorakhpur, Utkai, Gaurhati, North Bengal and Burdwan, to name a few.

As an expert geographer, Chatterjee had frequent occasions to travel to universities throughout the country, reviewing syllabuses, setting or moderating examinations, interviewing and selecting academic staff, and advising faculty or academic councils. Wherever he went, there would be former students awaiting him at the railway station or airport. He could thus retain a meaningful relationship with his students, scattered around the country. He also had an excellent rapport with scholars in sister disciplines, including anthropology, geology, Indology, ancient history, education, botany, zoology and commerce. Indeed, a noted anthropologist, Professor N.L. Bose, joined his department to teach human geography for a number of years before becoming director of the Anthropological Survey of India. During the period of Chatterjee's ascendancy, he certainly did have rivals – stalwarts in their own universities – like Kariyan of Madras, Dubey of Allahabad, Chibber of Banares and Rizvi of Aligarh, but nobody questioned Chatterjee's pre-eminent position in Indian geography, and everyone had been a guest at his home or office.

Chatterjee strode over India's geographical firmament like a colossus for nearly four decades. He almost single-handedly worked to upgrade the status of geography from the level of an elementary school subject to the highest plank in the university's academic hierarchy, and its social ranking from almost naught to the corridors of national planning. He was the first geographer to be dean of a faculty and a member of the senate and syndicate of a major Indian university. He was the first geographer to be appointed director of an Indian central government establishment.

Chatterjee also had important links and friendships in the international arena; the link with L. Dudley Stamp (1898–1966), Professor of Social Geography at the University of London, and Advisor to HM Government on Agriculture and Land Use, has already been mentioned. This association was important, as L.D. Stamp on numerous occasions sponsored Chatterjee for international recognition, and supported him strongly (and successfully) and his battle with the Survey of India to maintain the independent status of his National Atlas Organization. Mention must also be made of Professor A.M. Ryabchikov, former Dean of the Faculty of Geography at Moscow State University, who worked closely with Chatterjee on many IGU committees. Several Japanese geographers were Chatterjee's close personal friends, including Professor J. Yunekura of Hiroshima University. There were other important links with geographers in Sweden, France, Germany, the USA and the UK. His influence was thus widespread.

There is one significant contribution that Chatterjee made to the body politic of India and the Indian nation that remains relatively unknown: the name Meghalaya that he thoughtfully gave to his research area in Assam, comprising the Garo-Khasi-Jaintia Hills, while he was working France in the 1930s. He repeated the name as a physiographic division in the 1957 edition of India's national atlas, and also in his chapter on Physiography in the 1965 edition of the *Gazetteer of India.* (see above). It became the unanimous choice of the Indian parliament when the same region attained statehood. The State of Meghalaya found a place on the political map of India in 1972, a fitting memorial to the ingenuity of Chatterjee, and a unique example of the quiet influence of geography on politics.

Bibliography and Sources

1. SELECTED WORKS BY S.P. CHATTERJEE

* An asterisk indicates co-authorship or co-editorship

1928 'Geological reconnaissance in the Sohawal State, Central India', *Journal of Burma Research Society*, Vol. 18, No. 3.

1928 Physical geography of the Girnar Hills, Kathiawar', *Journal of Burma Research Society*, Vol. 18, No. 3.

1936 *Le Plateau de Meghalaya*, Les Presses Modernes, Paris.

1937 'The teaching of geography', *Calcutta Geographical Review*, Vol. 1, No. 2.

1939 'Geography of Puri and adjoining areas, Orrisa', *Calcutta Geographical Review*, Vol. 2, No. 2.

1939 *Poland and its Frontier*, Calcutta Geographical Society, Monograph 1.

1940 'Natural regions of Bengal', *Journal of Madras Geographical Society*, Vol. 15, No. 1, 1–18.

1940 *Turkey and its Problems*, Calcutta Geographical Society, Monograph 2.

1941 'The place of geography in national planning', *Calcutta Geographical Review*, Vol. 3, No. 1, 63–91.

1941 'Ice age in India', *Calcutta Geographical Review*, Vol. 3, No. 3.

1943 'Geographical interpretation of distribution of population in two typical districts of India', *Calcutta Geographical Review*, Vol. 5, No. 3.

1943 'Is Damodar taking to its former eastward course?', *Calcutta Geographical Review*, Vol. 5, No. 3.

1943 'World's cereal foods with special reference to India and Bengal', *Calcutta Geographical Review*, Vol. 5, No. 4.

1944 'The physiographic and economic bases of urbanization in the Gond and adjoining lands of the Central Provinces and Berar', *Calcutta Geographical Review*, Vol. 6, No. 1.

1947 'River problems in Bengal', *Punjab Geographical Review*, Lahore, Vol. 2, No. 1.

1947 *The Partition of Bengal, A Geographical Study with Maps*, Calcutta Geographical Society, Monograph 8.

1949 *Bengal in Maps*, Orient Longman Ltd, Calcutta.

1949 'Coastal geomorphology', *Calcutta Geographical Review*, Vol. 11, No. 1.

1949 *24 Parganas and Calcutta*, Calcutta City Book Co., Calcutta (Bengali).

1949 'Economic Survey of Europe', *Calcutta Geographical Review*, Vol. 11, No. 2.

1952 'Effect of partition on the economic geography of Bengal', in *Proceedings of the 17th International Geographical Congress*, Washington, 609–11.

1952 'Food shortage', *Geographical Review of India*, Vol. 14, No. 2.

1952 'Land utilization survey of Howrah district', *Geographical Review of India*, Vol.14, No. 3.

1952 'Indian and world food supply', *Geographical Review of India*, Vol. 14, No. 4.

1955 'Thematic maps of India, Second Geographical Congress of the Soviet Union', *Soviet Academy, Science Journal No. 6*, Moscow.

1956 'Pedogenesis in West Bengal – I', *Geographical Review of India*, Vol. 18, No. 3, 1–10.

1957 'Palaeogeography of India during Archaean times'. *The Observer*, Calcutta, 3.

1957 'A note for the Symposium on flood control', *Proceedings of the National Academy of Sciences (India)*, Allahabad, Vol. 26 (AV), 390–5.

1957 'The changing map of India', *Geographical Review of India*, Vol. 19, No. 2, 1–5.

1957 'Soil acidity study in the district of Howrah', *Geographical Review of India*, Vol. 19, No. 3, 6–9.

1957 'Pedogenesis in West Bengal – II', *Geographical Review of India*, Vol. 19, No. 4, 1–12.

1957 'Geographic regions of east Rajasthan Pathar', *Geographical Review of India*, Vol. 19, No. 4, 42–7.

1958 'The national atlas of India', *Geographical Review of India*, Vol. 20, Nos 1–4, 47–62.

1958 'Rumania through Indian eyes', *Bharat Rumania Maitri Samaj*, New Delhi, 17–25.

1960 'Hydrographic features of the Adi-Ganga Bhumi', *Geographical Review of India*, Vol. 22, No. 1, 56–9.

1960 'Presidential address at the 25th anniversary of the Geographical Society of India', *Geographical Review of India*, Vol. 22, No. 1, i-iv.

1960 'A comparative study of soils formed from different parent rocks (Bankura district, West Bengal)', *Geographical Review of India*, Vol. 22, No. 2, 44–8.*

1960 'Classification of the alluvial soils of Howrah, West Bengal', *Geographical Review of India*, Vol. 22, No. 3, 6–14.*

1960 'Presidential address', *Geographical Review of India*, Vol. 22, No. 2, 1–5.

1961 'Presidential address', *Geographical Review of India*, Vol. 23, No. 1, 1–11.

1961 'Pedogenesis in Rajmahal – I', *Geographical Review of India*, Vol. 23, No. 1, 12–17.*

1961 'Pedogenesis in Rajmahal – II', *Geographical Review of India*, Vol. 23, No. 2, 64–71.*

1961 'Pedogenesis in Rajmahal – III', *Geographical Review of India*, Vol. 23, No. 3.*

1961 'Floods in West Bengal', *Geographical Review of India*, Vol. 23, No. 3, 1–5.*

1961 'Hawaii – the 50th state of the USA', *The Observer*, Calcutta, 6 and 7.

1962 'Fluctuations of sea level around the coasts of India, with illustrations', *Zeitschrift für Geomorphologie*, Berlin, N.S., Supplement, Vol. 3, 48–56.

1962 'Snows and glaciers of India', *The Observer*, Calcutta, 8.

1962 'Planning for agricultural development in India', *National Geographer*, Vol. 5, 8–56.

1962 'Presidential address', *Geographical Review of India*, Vol. 24, No. 1, 1–6.

1962 'The nature of the clay minerals in the Rajmahal soil', *Geographical Review of India*, Vol. 24, No. 1.

1962 'Regional patterns of the density and distribution of population in India', *Geographical Review of India*, Vol. 24, No. 2, 1–28.

1962 *India's Village Population*, Khadi Gramodyog, Khadi and Village Industries Commission, Bombay.

1962 'An outline of the geography of Hawaii, USA, 10th Pacific Science Congress 1961, *Geographical Review of India*, Vol. 24, No. 4.

1963 *Fifty Years of Science in India, Progress of Geography in India During Past Fifty Years*, Indian Science Congress Association, Calcutta.

1963 'My recent visit to Hungary', *The National Geographical Journal of India*, Vol. 9, No. 2, 73–83.

1963 'Presidential address', *Geographical Review of India*, Vol. 25, No. 1, 1–10.

1963 'Fertility classification of some of soils of village Hat [Bansantapur (P.S. Arambag) and village Udaipur (P.S. Khanakul) District Hooghly] – a comparative study', *Geographical Review of India*, Vol. 25, No. 4.

1963 'National atlas of United Kingdom', *Geographical Review of India*, Vol. 25, No. 3, 194–5.

1963 'Our national and geographical atlases', *Geographical Magazine*, Bombay, 1–19.

1963 *Modern Cartography in India*, The Indian Archives, XV.

1964 'Settlement pattern in Puri-Chilka coastal track', *Geographical Review of India*, Vol. 26, No. 4.*

1964 'International Geographical Congress and International Geographical Union: Their history and development during Inter-war and post-war periods', *The Observer*, Calcutta, 1–8.

1965 'Le Regioni Dell' India (India With Maps)', *Società Geografica*, Rome, 1–26.

1965 'Physiography of India', *Gazetteer of India*, Government of India Publication, New Delhi, 1–65.

1965 'Presidential address', *Geographical Society of India*, Vol. 27, No. 1, 1–20.

1965 'Population mapping in India', *The Observer*, Calcutta, 1–3.

1965 'Region, regionalism and economic regionalization', *The Geographer*, Vol. 14.

1966 'Unión geográfica internacional: Conferencia regional latinoamericana', *Discursos y Conferencias*, Mexico, Vol. VII, 13–49.

1966 'Aspects of the study of regional geographical sketch, Rajony, India', *Acta Geologica et Geographica Universitatis Comenianae Geographica*, Bratislava, Vol. 6.

1966 'Our national and regional atlases', *Bombay Geographical Magazine*, Vol. XIV, No. 1.

1966 'Analysis of dependence of yield on soil nutrients of the soils growing potato in Arambagh Sub-Division', *Geographical Review of India*, Vol. 28, No. 2.

1966 'Human response to the Himalayan range', *The Observer*, Calcutta, 1–8.

1967 'Geomorphological mapping', in *Proceedings of the Seminar on Geomorphological Studies*, Sagar University, 177–87.

1967 'Geographical cartography', *Geographical Review of India*, Vol. 29, No. 1, 5–13.

1967 'Presidential address', *Geographical Review of India*, Vol. 29, No. 1, 1–13.

1967 'Rajmohal soils', *Geographical Review of India*, Vol. 29, No. 2.*

1967 'Study of organic matter in the forest soils in Darjeeling Himalayas', *Geographical Review of India*, Vol. 29, No. 3.*

1968 *National Atlas of India*, Fascicule volume (edited), Ministry of Education and Scientific Research, Calcutta.

1969 *Report of the Lower Damodar Valley Region*, Diagnostic Survey of the Damodar Valley Region, Damodar Valley Corporation, Calcutta.

1969 *Planning Atlas of the Damodar Valley Region*, Damodar Valley Corporation, Calcutta.

1969 'India: Land, people and economy', in *Enciclopedia di Tulli 1 Paesi del Mondo*, Vol. VI, Instituto Geografico de Agostini, Novara, 106–86.

1970 'Nature's balance', in *West Commemoration Volume*, Sagar University, 79–81.

1970 'India's present day relationship with the neighbouring countries', *Geographical Review of India*, Vol. 32, 1–20.

1970 'Land use map of West Bengal', *Journal of the Institute of Economic Geography*, Vol. 1, No. 1.

1970 *Meghalaya – Its Land and People. A Geographical Study*, Central Board of Irrigation and Power, Shillong.

1971 'Evolution of human settlements in India since the dawn of civilization', *Revue Roumaine de Géologie, Géophysique et Géographie*, Bucharest, Vol. 14, No. 1, 109–15.

1972 'Bhagirathi-Hooghly basin, India', in *Proceedings of the Interdisciplinary Symposium*, Calcutta, 19–24.

1972 'Towns and cities of India as evolved over time and space', *Journal of the Institute of Economic Geography*, Vol. 3, No. 1, 1–25.

1972 *Land use Map of India*, in two plates, scale 1 : 5 000 000, Instituto Geografico di Agostini, Novara.

1972 *Irrigation Atlas of India* (edited), Department of Science and Technology, Calcutta.

1973 'Asia', in *World Atlas of Agriculture*, General Section, Vol. 2, Instituto Geografico de Agostini, Novara, 1–557 (edited).

1973 *Progress in Geography, A Decade of Science in India (1963–72)*, Indian Science Congress Association, Calcutta.

1974 'River problems in India', in 'Proceedings of the Symposium on the Danube River', *Foldrajzi Ertesito*, Budapest, Vol. 23, 51–3.

1974 'Population dynamics', in *Seminar on Population Crises*, World Population year, 1–22.

1974 'A duna szimpozium erteklo osszefoglaeasa', *Foldrajzi Ertesito*, Budapest, Vol. 83, No. 1, 51–3.

1975 *An Introductory Regional Geography: West Bengal, Tripura, Bangladesh*, Orient Longman, Calcutta.

1976 *An Introductory Physical and Economic Geography: India*, Orient Longman, Calcutta.

1976 *An Introductory Regional Geography: India and the World*, Orient Longman, Calcutta.

1976 *India in Maps*, Kendall/Hunt Publishing Co., Iowa.

1977 *An Introductory Regional Geography: West Bengal*, Orient Longman, Calcutta.

1977 *An Introductory Regional Geography of India*, Orient Longman, Calcutta.

1978 *Elements of Physical Geography*, Junior College Geography Part I, Orient Longman, Calcutta.

1978 *Geography of Resources and Economic Activities*, Junior College Geography Part III, Orient Longman, Calcutta (reprinted in 1983).

1978 'Geography in the cultural and spiritual life of a nation', Sri Sathya Sai Summer Course in Indian Culture and Spirituality, Bangalore, 1–39.

1982 'Evolution of political history of India as influenced by geographical factors', *Geographical Review of India*, Vol. 44, No. 1, 1–18.

1983 'Evolution of human settlements in India since the dawn of civilization', in Mishra, R.P. (ed.), *Concepts and Approaches – Contributions to Indian Geography*, Heritage Publishers, New Delhi.

1983 *Geography for Secondary Students*, Part I, Orient Longman, Calcutta.

1984 *Economic Geography of Asia*, Allied Book Agency, Calcutta, Part I – *Asia as a Whole*, Part II – *South Asia*, Part III – *India*, Part IV – *South-West Asia, East Asia, Soviet Asia.*

1985 Latest venture of the Association of Japanese geographers, *Indian Journal of Landscape Systems and Ecological Studies*, Vol. 7, No. 2.

21st International Geographical Congress Publications, New Delhi, 1968

Towards Global Peace and Harmony: Rapproachment between Developing and Developed Countries, Presidential Address at the Congress.

La géographie régionale du plateau de Meghalaya', in *India Regional Studies*, National Committee of Geography, 218–44.

Progress in Geography in India (1964–68), Fifty years of Science in India, Indian Science Congress Association, Calcutta.

Physical Geography, Selected Papers, Vol. 1 (edited).*

Economic Geography, Selected Papers, Vol. 2 (edited).*

Population and Settlement Geography and Historical and Political Geography, Selected Papers, Vol. 3 (edited).*

Regional Geography and Cartography, Selected Papers, Vol. 4 (edited).*

Developing Countries of the World, Special Publication 1 (edited).

Congress Proceedings (edited).

2. *PUBLICATION DEDICATED TO PROFESSOR S.P. CHATTERJEE*

Choudhuri, M.R. *Essays in Geography*, The Geographical Society of India, Calcutta, 1965.

Sitanshu Mookerjee was Professor of Geography at the University of Calcutta, and Vice-Chancellor of the University of Kalyani, India.

Chronology

1903	Born in Bengal on 22 February
1926	Awarded MSc in Geology, Banares Hindu University
1928	Head of Department of Geology and Geography, University of Rangoon; becomes acquainted with L. Dudley Stamp
1932	Leaves University of Rangoon; proceeds to Europe for higher studies, initially at the Sorbonne, Paris, and later in London
1937	Returns to India; takes up position with Faculty of Education, University of Calcutta
1939	Introduces Honours programme in geography
1947	Independence and partition of India; Chatterjee prepares important monograph on partition of Bengal and attracts considerable attention; *Bengal in Maps* follows
1950	Starts work on Land Use Survey of Howrah District of West Bengal using methods of L.D. Stamp
1956	National Atlas Unit formed within geography department, University of Calcutta
1959	Murchison Award of the Royal Geographical Society
1963	Nehru Medal of the National Geographic Society of India
1967	Retired from University of Calcutta
1968	Organizes IGU Congress in Delhi; first Asian president of IGU
1985	Decorated by president of India
1986	Dies in Calcutta on 27 February

Frederic Edward Clements

1874–1945

Patrick H. Armstrong
and Geoffrey J. Martin

Frederic Clements was one of the founders of modern ecology, a discipline that has profoundly influenced the scope and direction of geographical enquiry, and greatly influenced the manner in which many geography students were taught for much of the twentieth century. Originally given to the study of phytogeography, Clements' name will always be linked with the notions of the plant community as an integrated, holistic entity, and those of succession and climax.

1. Education, Life and Work

Frederic Clements' grandfather, George William Clements, was born in Somerset, England, in 1809. He married Harriet Richards in 1832 and sailed to America ten years later. They settled in Marcellus, New York, where they had a son, Ephraim G. Clements, born in 1843. Ephraim married Mary A. Scoggin and moved to Lincoln, Nebraska, a pioneer village, where he established a photographer's studio. Frederic was born on 16 September 1874.

He was raised 'just down the street' from the young University of Nebraska, where he began his undergraduate work at the age of 16 years. While attending the university, he served as cadet captain under the command of Lieutenant John J. Pershing. Associated with him were Willa Cather, Homer L. Shantz, Alvin Johnson and Roscoe Pound. Although Clements was described as prim, ascetic and humourless, his brilliance was recognized by Charles E. Bessey (1845–1915), himself a student of Asa Gray and at that time head of the infant Department of Botany. While an undergraduate, he undertook field and laboratory work on the taxonomy of flowering plants and fungi. Then Clements was put to work with another protégé, Roscoe Pound (1870–1964), to compile an inventory of the original vegetation types of the frontier state in 'a race ... against time, homesteaders, and the westward march' (Worster, 1985). At this time he became a member of the Botanical Seminar, which was an informal group of the best minds bringing science to the prairies only recently ploughed. This group provided a rigorous schooling long remembered by its participants. Clements graduated with a BSc degree from the University of Nebraska in 1894, and then earned AM and PhD degrees in 1896 and 1898 respectively. (The LLD degree was conferred upon him in 1940.) Clements' two advanced degrees bespoke an interest in morphology, taxonomy, phylogeny and ecology which was to stay with him throughout his life.

In 1894 he was appointed instructor and laboratory assistant in the Department of Botany of the University of Nebraska; later he was promoted to the rank of associate professor of botany, and then professor of plant physiology (1905). Yet is was in 1894–5 that American plant geography received a fillip. Bessey had developed a fine library, and a respect for laboratory instruction in his newly emerging field of study. In 1896 Bessey asked Pound to review Oscar Drude's (1852–1933) *Pflanzengeographie von Deutschland*. 'At the time, after reading Drude's book, I got the idea of a Phytogeography of Nebraska' (Pound, 1954). Clements had written 'the Histogenesis of the Caryophyllales' (as an AM thesis) and was now writing 'The Phytogeography of Nebraska' as a doctoral thesis. Clements meanwhile read Drude. Since Pound and Clements were working so closely together, this doctoral thesis was elaborated and published with joint authorship. This was Clements' first book; a second edition was published in 1901.

In the period 1896–8, Clements and Pound published seven papers between them (Pound, 1954). In essence, what they attempted for the first time was the application of quantitative methods to ecology. In place of description of vegetation types, they carefully analysed the composition of prairie communities, showing the real nature of the imperceptible changes that occur as one moves over the plains. Here was an opportunity for a university adjacent to a newly opened prairie province. Interrupting the valuable confluence of these two minds was the departure of Roscoe Pound into the field of law; in 1903 he became Dean of Law at Nebraska. (In 1907 he took a post as professor of law at Northwestern University, eventually becoming Dean of the Law School at Harvard University, and a noted writer on jurisprudence.)

The year 1898 also witnessed another major contribution to the development of plant ecology in the USA; H.C. Cowles, the first doctoral candidate of J.M. Coulter at the University of Chicago, contributed a significant thesis on the dynamics of sand dune vegetation.

In that same year, Edith G. Schwartz was graduated from the University of Nebraska, where later she earned a doctorate in botany. Clements married her that same year. She was a talented artist, botanist and author and shared his travail until his death in 1945. Together they produced *Rocky Mountain Flowers* (1914) and *Flower Families and Ancestors* (1928).

Clements departed from Nebraska in 1907 to become Professor and Head of the Botany Department at the University of Minnesota, a post he held until 1917, when he became a research associate in ecology with the Carnegie Institution in Washington (the organization that had published his best-known work *Plant Succession* in 1916).

At school he had obtained a grounding in Latin and literature; as an adult he began the study of several languages, including Polish, in order to read Rostafinski's book on slime-moulds. Both he and his wife (who had taught German before specializing in botany) enrolled in a course in comparative philology. This would appear to have been of value to him when he wrote 'Greek and Latin in biological nomenclature' in 1902.

Clements benefited from a reading of Warming's *Plant Ecology*, Schimper's *Physiological Plant Geography*, and Merriam's *Life Zones and Crop Zones of the United States*. Then he wrote *Research Methods in Ecology* (1905), which was a major contribution to the fledgling field of ecology. Aided by his wife, Clements created an alpine laboratory in the Rocky Mountains, on the slope of Engelmann Canyon below Pikes Peak. Of that locality, Clements calculated 'in the distance of nine miles from Manitou at 6,000 feet to the summit at 14,000 feet, there were as many different climates as there are from the Gulf of Mexico to the Arctic Circle'. His work for many summers yet to come was accomplished here; eventually the station was taken over by the Carnegie Institute and properly funded.

Linkages between geography and the emerging but ill-defined field of ecology were forged. Clements had taken membership in the American Geographical Society in 1900, and in 1904 was listed as one of the founding (48) members of the Association of American Geographers. Interestingly, Clements was nominated by H.C. Cowles and W.M. Davis, both of the organizing committee. Davis wrote to Secretary Brigham of Clements on 13 January 1905 that 'he is a valuable man'. Clements had a paper read by title at the initial meeting in Philadelphia, 1904, 'The interaction of physiography and plant succession in the Rocky Mountains', which was at least in part the product of his experience in the 'alpine laboratory'. He did not attend the meeting, or those that followed: Clements was some 1400 miles removed from the east coast where the meetings were invariably held.

There was nevertheless a community of ecologically minded scholars within this association which included Charles C. Adams, Henry C. Cowles, Curtis F. Marbut, Charles Merriam and Leonhard Stejneger. It was not long before other ecology-minded members were brought into the association: the following year J.A. Allen, Arnold E. Ortmann and Edgar N. Transeau were added to the membership. Clements, however, was active in a number of other societies. He took membership in the American Academy of Arts and Sciences in 1896, was made a Fellow in 1898, and was a councillor in 1905 and 1909; he was also a member of the Botanical Society of America from 1898, vice-president in 1905, and councillor from 1906 to 1910; and he was director of the Society of American Naturalists in 1905 (holding also numerous other memberships and offices).

Humboldt's writings on vegetation seem to have had some indirect influence on Clements through the German school of *Pflanzengeographie*. In the summer of 1911, Clements was able to join Tansley, Drude and a number of other European ecologists on the International Phytogeographical Excursion, visiting a large number of field sites in East Anglia, northern England, the West Country, Scotland and Ireland; moorland, heath, wetland and woodland communities were examined. In 1912, a Second International Phytogeographical Excursion was organized, crossing the USA from New York to California. During this time, Clements and his American colleagues were able to exchange thoughts with their European colleagues, who included Tansley (England), Paulsen (Denmark), Stomps (The Netherlands), Skottsberg (Sweden), Engler and Tubeuf (Germany) and yet others. The American ecologists showed their European friends the plant communities of the eastern forests, the prairie grasslands, and the redwoods and chaparral of California.

These excursions had the effect, as one participant on the first journey put it, of 'internationalizing for all time' the subject of phytogeography. The geographical approach had been of value to Clements, and his friendship with geographers had been increasing. He also wrote extensively on the effects of climate on plants and vegetation (a paper on 'Drought periods and climatic cycles' was published in

Ecology in 1921), and he took up membership of the American Meteorological Society.

In 1917 he left Minnesota to take a post as research associate of the Carnegie Institution of Washington, though he was posted in Tucson, Arizona until 1925. The Carnegie Desert Laboratory had been established in Tucson 1902. Several climatological stations had been established on nearby mountains, and camels were used for local transportation. This haven for eremologists attracted Douglas T. Macdougal, Godfrey Sykes, Forrest Shreve, William A. Cannon, Robert H. Forbes and William L. Tower, and others. Each of these workers had a keen interest in some facet of the ecological undertaking. Many of them were keenly interested in botany. Clements found himself in a stimulating physical and intellectual environment. A well-known photograph taken at this time (and later reproduced in a memorial typifying him as a muddy-boots botanist) shows him clothed in collar, tie and smart jacket, but wearing stout boots and broad-brimmed hat, surrounded by Cactaceae.

During the summers, he and his wife continued to give their time to the alpine laboratory and Pike's Peak, Colorado. After his early work on grasslands, Clements began to write of the forests of the Rockies. A publication entitled 'Life-history of lodgepole burn forests' represented an appreciable shift in emphasis to a more applied focus (it was written for the US Forest Service), as well as a movement from the prairies to the coniferous forests. Indeed, 'applied ecology' was an important theme in his work.

In 1917 Clements was given a special assignment connected with the war effort. He was to study the western and north-western ranges in order to determine how effectively sheep- and cattle-men were practising sensible grazing methods. This meant travelling many hundreds of miles in 'Billy Buick', Mrs Clements at the wheel. Now working for the Carnegie, Clements had the botanist Hall of Berkeley appointed to his staff and at once assigned to the task of searching for rubber that might be extracted from plants growing in the sagebrush country. The investigation concerning grazing continued for several more weeks, taking the Clementses deep into Texas, then back to the alpine laboratory. Later, he contributed to the work of the Soil Conservation Service, as well as organizations involved in forestry and nature conservation.

In 1925 he and his wife and colleague, Edith, moved to the coastal laboratory of the Carnegie Institution at Santa Barbara, California. As a result of this variety of locale and experience, his knowledge of the plants and the plant communities of the USA was unrivalled.

He was a profoundly religious man (he had a puritanical streak, and is said to have detested smoking and drinking), and this may have had implications for his work. Some have said that he was also arrogant, but a letter from his wife written two years after his death is interesting:

> his 'arrogance' was purely intellectual. ... Temperamentally, he was extremely shy and self-deprecatory. Haven't you noticed that many painfully shy people give the impression of being haughty and stand-offish? It is a sort of protective mechanism, I presume. Of course, too, his very sensitively organized body kept him from joining in the usual associations of more robust constitutions, and that was often misunderstood and misinterpreted. (Mrs Edith Clements to A.G. Tansley, 15 July 1947)

He was an individualist in other ways too. Perhaps what was not widely known had a greater effect upon his person and his manner. Clements was a diabetic before the advent of insulin. Only most exacting regimes of diet, work and sleep imposed by his wife allowed him to contribute so much to his chosen field.

2. Scientific Ideas and Geographical Thought

The most significant contribution that Clements made to plant ecology was the pioneering work *Plant Succession; an Analysis of the Development of Vegetation* (1916). This work was revisited in 1928 when his *Plant Succession* and *Plant Indicators* (1920) were combined and condensed to form *Plant Succession and Indicators*.

The concept of 'succession' was not minted by Clements, as he demonstrates in an historical review of the literature. However, he returned time and again to the term and the theme, clothed its modes and nuances in a series of technical terms derived from his horde of field experiences, and then presented the whole to an international audience.

'Succession' was a pioneer theme in the context of an emerging ecological undertaking. In his 1916 work he quoted and evaluated evidence provided by naturalists in a number of countries since the seventeenth century, and then recounted his own fieldwork in the mid-west. From a careful weighing of the evidence, coupled with his own experience, Clements developed an elegant conceptual framework for the analysis of plant communities.

His essential thesis was that a plant community, developing in an area previously devoid of life – the *initial stage*, passed through a *pioneer* stage (made up of the first plants to colonize the bare area) and then through a set of *seral* communities, to a *climax*, a mature, stable assemblage of plants that underwent little further change, and that was in equilibrium with the climate of the region. Although there might be differences in the pioneer communities and the seral stages within a region, depending upon the nature of the original substratum (bare rock, volcanic ash, sand dunes, water), the climaxes would be much the same.

A distinctive component of Clements' thinking was that the plant community was a sort of superorganism:

> As an organism the formation arises, grows, matures and dies ... Furthermore, each climax formation is able to reproduce itself, repeating with essential fidelity the stages of its development. The life-history of a formation is a complex but definite process, comparable in its chief features to the life-history of an individual plant. (Clements, 1916, p. 3)

An extension of this idea is that the sequences of change in succession not only recapitulated the life-cycle of the individual organism, but also the evolution of plants from 'lower to higher life forms'. (It is recorded that Pound and Clements had together studied Herbert Spencer's *Principles of Biology*, published in 1874; this book emphasized the idea that human society was a superorganism, but, according to Pound, writing over half a century later, at the time they got 'nothing out of the book'.)

Perhaps to some extent because of his religious views, Clements was not an uncritical follower of Darwin. Tansley (see below), in his obituary of Clements, defined him as a Lamarckian, and somewhat out of touch with 'the results of modern genetical research'. There are surprisingly few direct references to Darwin in his writing, though he knew the man's work and had great respect for it, as is shown in his 'Darwin's influence upon plant geography and ecology' (1909). Clements' work was part of the broad sweep of the Darwinian revolution in science, and it certainly emphasizes gradual change – 'the present is the natural and inevitable outcome of the past' (Clements, 1916, p. 279). In his abundant use of metaphor and allegory, in his certainty of the directionality of successional change, and in his insistence on the importance of what modern writers might call 'feedback' between organism and environment, Clements' thought closely resembled that of Darwin:

> In this development, habitat and population act and react upon each other, alternating as cause and effect until a state of equilibrium is reached. The factors of the habitat are the causes of the responses of the community, and these are the causes of growth and development, and hence of structure, essentially as in the individual. (Clements, 1916, p. 6)

While he acknowledged the importance of Charles Lyell (1797–1875), his debt to Darwin was indirect, for he used the ideas of T.C. Chamberlin (1843–1928), R.D Salisbury (1858–1922) and W.M. Davis (1850–1934), seeing analogies between the development of landforms and that of plant communities. But he also emphasized competition, 'the universal characteristic of all plant communities', and long-distance dispersal, 'agents which carry migrules ... [include] wind, water, glaciers, man and animals'; both are profoundly Darwinian concepts.

Another of the newly emergent geographer-ecologists was Henry C. Cowles of the University of Chicago. He was contemporaneous with Clements, and their careers were strikingly parallel: Cowles obtained his doctorate from Chicago in 1898, taught courses in plant ecology, and also became a founding member of the Association of American Geographers (AAG). He contributed regularly to this infant association and became its president in 1910. His work seems to have made its way through J. Paul Goode to H.H. Barrows, whose AAG presidential address 'Geography as human ecology' was one of the most quoted in association history. (See 'John Paul Goode' Volume 8, this series, G. Martin.) Interestingly, Park, Burgess, and McKenzie exploited the ecological theme in a sociological context also at the University of Chicago.

The two centres of ecological advance in North America came from Chicago and Nebraska. Arguably, Chicago derived its point of view from de Candolle, Darwin and Warming, while Nebraska was fed by Humboldt, Grisebach and Drude; at least this is the viewpoint of Tobey. In any case, Cowles and Clements were developing an ecological plant geography of much significance to geography. While for Clements succession was unidirectional, Cowles believed that the process could be retrogressive as well as progressive. This was a major discord between two of the chief proponents of a subject matter in search of a discipline. It did not lead to squabbling, in part doubtless to the tenor of the times, and in part to the larger problem of establishing a niche in the curriculum for things ecological. In any case, the presence and contribution of Cowles constituted an influence on Clements' science.

Another important influence on his scientific life was that of the British ecologist (Sir) Arthur Tansley (1871–1955). They met on the International Excursions, and corresponded throughout life. (See 'Arthur George Tansley', Volume 13, this series, P.H. Armstrong.) Frequently, they exchanged ideas with each other before they ventured into publication. A single letter extract must serve to show the manner in which Clements introduced newly minted concepts and terms to his English friend before publishing them in *Plant Succession*:

> I am very glad to have your letter, and to know that you approve the developmental concept of the formation so far as I made it clear. I am getting more and more enthusiastic about it the longer I work it over and try it on. It is just what I was after in the original distinction between primary and secondary successions, but the organic relations were in no wise as clear as they are now .... I have been using the term 'sere'

> for a concrete succession, retaining 'succession' for the general field or process. In addition, there can be distinguished a 'clisere', a succession which is terminated by the replacement of the climax by one or more subsequent climaxes. There is also 'consere' or series of seres, the 'eosere' or the sum total of succession within the flora of a distinct period geologically, and the 'geosere', the total of all successional movements from the first plant population of the globe to the present. (Clements to A.G. Tansley, 29 May 1914)

Possibly Clements thought that Tansley and he were more 'nearly in accord' than they actually were, for Tansley never liked either the superorganism idea or some of the plethora of terms that Clements coined. Perhaps Clements was just ahead of his time, for there are phrases in the above that have a distinctly modern ring.

3. Influence and Spread of Ideas

Clements' influence on ecology was profound. During his lifetime, despite his occasionally rather difficult personality, he had many devotees. The inherent elegance of his carefully developed concepts of succession and climax, together with analogous 'gradualist' or 'uniformitarian' concepts in biology, geology, geomorphology and other sciences, ensured that generations of students (of physical geography as well as botany) learnt his terminology, and attempted to apply at least some of his ideas in the field. Most ecology and biogeography books published over the last three-quarters of a century have at least noticed the concept of succession, and many have dwelt on it. The idea has been applied to lakes and wetlands, glacial moraines, volcanoes, 'oldfield' communities (abandoned farmland – secondary succession), coral atolls, rubbish dumps, and sites affected by the testing of nuclear weapons. At the microhabitat level, the succession in decaying timber and carrion has been described. In urban geography, the idea of successional change has been applied to describe change in the ethnic and social composition of human communities in towns and cities.

In the wake of the depression and drought of the 1930s which devastated the grasslands of North America, Clements was frequently consulted on matters concerning land use policy and immediate problems concerning reclamation and restoration of western lands. It was in this context that he integrated humankind into this landscape to produce a very total ecology. His 'Environment and life in the Great Plains' and 'Climatic cycles and human populations in the Great Plains' were part of this thinking.

Clements' writings possibly may have had, via the teaching of Cowles, an indirect effect on Carl Sauer (1889–1975) when the latter was developing his ideas on the cultural landscape. In a 1925 paper, for instance, Sauer wrote of the 'organic' nature of landscape and of a 'developmental series' of landscapes, 'ranging from incipient to final or completed stage' (although the ideas of Davis and Salisbury were also important).

The cultural landscape notion received a boost when Derwent Whittlesey (1890–1956), in his essay on sequent occupance, wrote that the 'human occupance of an area, like other biotic phenomena, carried within itself the seeds of its own transformation' and that 'the analogy between sequent occupance in chorology and plant succession in botany will be apparent to all'.

Meanwhile, changes were taking place in the nature of the ecological succession concept itself. The idea of a plant community as a kind of superorganism never caught on (although Jan Christian Smuts – South African soldier, statesman, and mystic – acknowledged Clements in his influential *Holism and Evolution*, published in 1926, and the notion is not without resemblances to the modern 'Gaia' hypothesis). Colinveau referred to 'the heresy of the superorganism', perhaps an oblique dig at Clements' puritan zealotry, as well as his ecology. And many ecologists (Tansley among them) were not happy with the idea that for any area there was one regional climax to which all seral communities trended. A notion of *polyclimax* displaced the idea of unitary climax; this held that the precise form of the climax depended on the nature of the soil and environment at each place. Others were critical of the idea of a mature, stable climax in equilibrium with the climate; climates are continually changing, organisms are continually evolving, and on a local scale renewal and replacement are always taking place, as, for example, in a forest, where old trees die, collapse and are replaced by growing saplings.

There were other developments; E.P. Odum, in the 1960s, attempted to combine the succession principle with Tansley's ecosystem concept, considering, for example, the way in which energy flowed through, nutrients circulated around, and biomass accumulated within a community as it moved through seral stages towards climax.

P.A. Colinveau in 1973 summarized the situation:

> All the main phenomena of ecology seemed linked to this central phenomenon of succession. Ecological dominance; commonness and rarity; the regulation of number; the distinctness of species; the tuning of lives to share space, nutrients and energy; all must reflect that curious orderliness which seemed inseparable from succession. (*Introduction to Ecology*, 1973, p. 550)

But it was all too perfect and too deterministic, and there were many attempts at refinement, and to make the notion more explanatory. Connell and Statyer (1977), for example, investigated the changes in the species composition of plant communities in terms of three models of succession – the *facilitation, tolerance* and *inhibition* models.

In the facilitation model, later successional species can only become established after early colonizers have sufficiently modified the environment. In the tolerance model, the effects on the environment of early colonists neither increase nor decrease the rates of recruitment of later species: later appearances in a sere are just a manifestation of growth over time. With the inhibition model, the first invaders secure space and other resources, inhibiting the growth of other colonists, and indeed suppressing those that are present.

Careful observation suggested that reality did not always conform to the dictates of theory, however elegant. D. Walker found in a study of infilled peat hollows in Britain that a careful study of their stratigraphy did not show a sequence of deposits that confirmed the idealized progression from open water, through swamp communities, to woodland: things were much more 'messy'. Similar conclusions were reached by those who monitored the invasion of vegetation and animal communities following the Mount St Helens eruption in 1980.

But although some authorities have suggested the abandonment of the succession concept (e.g. C.J. Burrows in *Processes of Vegetation Change*, Unwin Hyman, London, 1990), it obstinately remains in the psyche of ecologists, especially as a pedagogic device.

Almost forgotten now is the fact that Clements' wife Edith wrote a doctoral dissertation (University of Nebraska) dealing with anatomical changes in leaves which were the product of environment. This came at a time when the concept of variation was being investigated by ecologists, geneticists, geographers and yet others. Clements became very interested in the plasticity of plants under differing circumstances. He experimented with transplants and observed their responses in structure to changed environment: hence *Experimental Pollination; an Outline of Ecology of Flowers and Insects* (with Frances L. Long) (1923), and *Experimental Vegetation: the Relation of Climaxes to Climates* (with John E. Weaver) (1924). Stanley Cain recognized the significance of this work in the acknowledgements to his book *Foundations of Plant Geography*.

Despite the close and very critical examination to which his ideas have been subjected, his legacy is enormous. Several organizations and scholarships that Frederic Clements helped to found and support (or were founded in his memory) have become of world importance. Many ecologists, geographers and other scientists have been attracted to their subject through the appeal of diagrams showing the steady progression from sand dune through scrubland to woodland, or from pond and reedswamp to tangled woodland. His concepts help to reveal something of the hidden mechanisms so essential to thoughtful planetary stewardship.

His finest memorials of an immediate sort were surely the two works left behind by his wife Edith, a woman of the largest capacity, and herself a force in the world of botany. In 1949 she joined with B.W. Allred to compose and edit *Dynamics of Vegetation: Selections From the Writings of Frederic E. Clements* (1949), and in 1960 she wrote, of their lives together, *Adventures in Ecology: Half a Million Miles from Mud to Macadam.*

Bibliography and Sources

1. SELECT BIBLIOGRAPHY OF WORKS BY FREDERIC CLEMENTS

1897 'Peculiar zonal formations of the Great Plains', *American Naturalist*, Vol. 31, 968–71.

1900 (with R. Pound) *The Phytogeography of Nebraska*, University Publishing Company, Lincoln, Nebraska.

1902 'System of nomenclature for phytogeography', *Engler Jahrb.*, b, 1.

1902 *Greek and Latin in Biological Nomenclature*, University Studies, Lincoln, Nebraska, Vol. 3, No. 1, pp. 1–86.

1905 *Research Methods in Ecology*, University Publishing Company, Lincoln, Nebraska.

1907 *Plant Physiology and Ecology*, H. Holt & Company, New York.

1909 *The Genera of Fungi*, The H.W. Wilson Company, Minneapolis.

1910 'The life history of lodgepole burn forests', *US Forest Service Bulletin*, 79.

1912 'Phytogeographical excursion in the British Isles. VIII, Some interpretations and reflections', *New Phytologist*, Vol. 11, 177–9.

1916 *Plant Succession*, Carnegie Institution, Washington.

1920 *Plant Indicators*, Carnegie Institution, Washington.

1928 *Plant Succession and Indicators*, The H.W. Wilson Company, New York City.

1929 (with J.E. Weaver and H.C. Hanson) *Plant Competition: an Analysis of Community Functions*, Carnegie Institution, Washington.

1934 'The relict method dynamic ecology', *Journal of Ecology*, Vol. 22, 39–68.

1936 'The nature and structure of the climax', *Journal of Ecology*, vol. 24, 252–84.

1936 'Environment and life in the Great Plains', *Carnegie Institution Supplementary Publication*, 24.

1938 (with J.E. Weaver) *Plant Ecology*, McGraw-Hill Book Company, New York and London.

1938 'Climatic cycles and human populations in the Great Plains', *Scientific Monthly*, vol. XLVII, No. 3, 193–210.

2. *OBITUARIES AND OTHER BIOGRAPHICAL ACCOUNTS OF FREDERIC E. CLEMENTS*

Shantz, H.L., 'Frederic Edward Clements (1874–1945)', *Ecology*, Vol. 26, No. 4 (1945), 317–19.

Tansley, A.G., 'Frederic Edward Clements, 1874–1945', *Journal of Ecology*, Vol. 34 (1946), 194–6.

Allred, B.W. and Clements, E.S. (eds), *Dynamics of Vegetation: Selections from the Writings of Frederic E. Clements*, H.W. Wilson, New York, 1946.

Pool, R.J., 'Frederic Edward Clements', *Ecology*, Vol. 35, No. 2 (1954), 109–12. (This contains a substantial, but not exhaustive, listing of his publications.)

Pound, R., 'Frederic E. Clements as I knew him', *Ecology*, Vol. 35 (1954), 112–13.

Phillips, J., 'A tribute to Frederic E. Clements and his concepts in ecology', *Ecology*, Vol. 35 (1954), 114–15.

Several of the above include brief critical appraisals of Clements' work.

3. *MATERIALS ON THE HISTORY OF ECOLOGY*

The history of the discipline has its own extensive literature. The authors found the following of value:

Worster, D., *Nature's Economy: a History of Ecological Ideas*, Cambridge University Press, Cambridge, 1985 (Originally published in the USA in 1977 by Sierra Club Books.)

Sheail, J., *Seventy-five Years in Ecology*. The British Ecological Society, Blackwell Scientific, Oxford, 1987.

A useful critique of the concepts of succession and climax is to be found in:

Connell, J. and Slatyer, R.O., 'Mechanisms of succession in natural communities and their role in community stability and organisation', *American Naturalist*, Vol. 111 (1977), 1119–44.

4. ARCHIVAL SOURCES

Material relating to the formation of the Association of American Geographers is to be found in the Association Archives, the American Philosophical Society, Philadelphia.

The University of Nebraska Archives, the University of Minnesota Archives and the Carnegie Institute each contain archival and other materials relating to F.E. Clements. Other material is held in the Frederic Clements Memorial at Texas A & M University, College Station, Texas. Clements' letters are, of course, widely dispersed; his correspondence with (Sir) Arthur Tansley is particularly important, and several of his letters (and the occasional letter from his widow) are to be found amongst the Tansley Papers, in the Botany School at the University of Cambridge.

Patrick H. Armstrong is Senior Lecturer in Geography at the University of Western Australia, Nedlands, Perth. Geoffrey J. Martin is Connecticut State University Professor and Professor of Geography at Southern Connecticut State University, New Haven, Connecticut, USA.

Chronology

1874	Born at Lincoln, Nebraska, 16 September.
1894	BSc degree, University of Nebraska; appointed instructor in the Department of Botany, University of Nebraska
1898	PhD degree, 'The phytogeography of Nebraska'; married Edith G. Schwartz
1904	Founding member, Association of American Geographers
1905	*Research Methods in Ecology*; appointed Professor of Plant Physiology, University of Nebraska
1907	Professor and Head of the Department of Botany, University of Minnesota
1911	First International Phytogeographical Excursion (British Isles)
1912	Second International Phytogeographical Excursion (USA)
1916	*Plant Succession*
1917	Research associate, Carnegie Institution; posted to Tucson, Arizona
1925	Posted to Santa Barbara, California
1928	*Plant Succession and Indicators*
1936	*Environment and Life in the Great Plains*
1945	Died, Santa Barbara, California on 26 July.

Roger Dion

1896–1981

Numa Broc

Collège de France

Considered one of the best French specialists in regional and rural geography in the earlier part of his career, Roger Dion later turned to historical geography.

1. Education and Career

Roger Dion was born on 28 October 1896, in Argenton-sur-Creuse, where his father was bailiff. After a brief stay in Paris (1904–7), he spent most of his adolescence in Blois, where he attended high school and obtained his baccalaureate in philosophy in 1913. He demonstrated an early interest in geography and was most sensitive to the beautiful landscapes of the Loire Valley and of Brittany where he spent his vacations. In 1910, the *Tableau de la géographie de la France* by Vidal de la Blache (*Panorama of the Geography of France*) became his bedside reading. Intending to become a teacher, he prepared for the entrance examination to the Ecole Normale at Louis le Grand High School in Paris (1913–15). However, he was called to the military in 1916, and served as an artilleryman for the rest of the First World War. In 1918 he began study towards a BA at the Ecole Normale and then an MA under the direction of Albert Demangeon. His thesis on the Blois section of the Loire Valley is entitled 'Le Val Blésois' ('The Blois valley', 1921). He obtained his aggregation with high honours in the same year. Instead of entering upon a teaching career, he became a member of the Thiers Foundation, where he started working on his doctorate (1922–4). He then returned to the Ecole Normale Supérieure as tutor and as vice-principal. He actively pursued research in the field, developed friendships with geographers of his age such as Pierre Deffontaines and taught at the Sorbonne from 1927 to 1932.

On 19 December 1933 at the Sorbonne, Dion defended his thesis *Le Val de Loire, étude de géographie régionale* (*The Loire valley, a study in regional geography*) in front of a jury presided over by the then master of regional geography in France, Albert Demangeon. His thesis was unanimously recognized as a masterpiece, both from a scientific point of view and for its original form. Dion was then named assistant professor and later full professor at the Faculty of Letters in Lille, where he taught from 1934 to 1945. There he discovered the problems faced by a large industrial and urbanized border region.

In 1945 he became Professor of Economic and Political Geography at the Sorbonne, where the memory of Vidal de la Blache and Demangeon was still strong. He also taught the geography of France at the Ecole Normale Supérieure, after a year spent at the University of São Paulo (1946–7). Roger Dion was finally named Professor of the Historical Geography of France at the Collège de France. He kept the appointment for 20 years from 1948 to 1968. He then retired without abandoning his research. Roger Dion died in Paris on 19 September 1981.

2. Scientific Ideas and Geographical Thought

Three major themes characterize Dion's work: rural landscapes, the history of vine-growing in France, and historical geography.

RURAL LANDSCAPES

Ever since his thesis on the *Val de Loire*, Dion had concentrated on agricultural modes of exploitation and the landscapes they create. He demonstrated how, in the Loire Valley, two types of rural economy and two types of landscape competed: the open field and the mixed woodland and pasture. Avoiding the traditional opposition of physical and human geography, he emphasized the role of history which, more than geology, explained the diversity of Loire landscapes. Broadening his attention to the whole of France, Dion published in 1934 a brief 'Essai sur la formation du paysage rural français' (Essay on the creation of the French rural landscape) which was an extension of *Caractères originaux de l'histoire rurale française* (*Original characteristics of French rural history*, 1931) published by the historian Marc Bloch. Picking up the old debate which had pitted the advocates of geological determinism (Risler) against those of ethnographic determinism (Meitzen), he proved that the opposition between mixed woodland and pasture and open field was ancient. It reflected the very different agricultural civilizations that had been diffused by the Celts, the Germans and the Romans. Following the ideas of the German Meitzen, he traced a line on a map of France from the estuary of the Seine to Geneva. This line separated northern France from southern France, the France of open fields and clustered settlement from the France of mixed woodland and pastures with dispersed settlement (*bocage*).

After 1934, Dion's concepts became more focused. In the 1939 article 'Les Principaux Types du paysage rural' ('Main types of rural landscape'), he proposed a four-tiered classification: open field, mixed woodland and pasture, villages of the centre and west, and southern fields. In 'La Part de la géographie et celle de l'histoire dans l'explication de l'habitat rural du bassin parisien' ('The role of geography and of history in explaining rural settlement in the Paris Basin') (1946), he underlined the importance of the medieval historical period on economic and demographic factors. The clearing efforts of the eleventh and twelfth centuries were responsible for contemporary rural landscapes in the Paris Basin. He asserted that 'human will' prevailed over geographic and ethnographic determinism. In his 'Réflexions de méthode à propos de *La Grande Limagne* de Max Derruau' ('Reflections on methodology relative to Max Derruau's *The Great Limagne*)

(1951), Dion showed that one cannot ignore recent economic and social history (sixteenth through eighteenth centuries) when trying to understand the latest evolution of our rural landscapes.

THE HISTORY OF VINE-GROWING IN FRANCE

How does one explain the spatial distribution of vineyards in France? As early as 1933, Dion asked that question about the Loire Valley. Though he did not neglect natural conditions (soils, climate, etc.), he favoured the role of historical factors: the importance of shipping on the Loire facilitating wine transport, and the role of princes and of the urban middle classes who encouraged the multiplication of vineyards. Following the publication of several monographs on the main French vineyards in the Middle Ages, Dion wrote in 1959 his monumental *Histoire de la vigne et du vin en France des origines au XIXème siècle* (*The history of vines and wine in France from their origins to the nineteenth century*). He showed how the Romans developed vine-growing along the major rivers. He mostly demonstrated how vine-growing spread over the whole country during medieval times, independently of the kind of weather or soil. A rich and prestigious kind of cultivation which developed around castles and cities, vine-growing was also closely linked to the Christian religion. The bishop was often 'the first vine-grower of the city'. For the modern period (seventeenth to eighteenth centuries), Dion was more interested in the development of 'popular' vineyards, producing wine for everyday consumption (such as the Orleans or Parisian vineyards) than the promotion of quality vineyards (such as those of Bordeaux or Bourgogne). He barely mentioned the Mediterranean vineyards of very recent vintage (nineteenth century). 'France', concluded Dion, 'is a country where the history of vine growing informs that of a whole people'; Dion's historicizing theses about the origin of French vineyards troubled many geographers like Max Sorre, who were more used to ecological concepts. These original thoughts, which could sometimes be, arguably, dismissed, may have contributed to Dion's distance from traditional geography.

HISTORICAL GEOGRAPHY

Using history to explain geographic phenomena naturally led Dion to historical geography. In *Les Frontières de la France* (*France's borders*), (1947), he returned again to the Middle Ages to demonstrate the permanency of certain interior boundaries. He showed that Gallo-Roman cities and Carolingian counties often had geometric borders. Feudality introduced 'an imperfection contrary to Roman traditions', in its territorial divisions. His work also included a discussion of the concept of 'natural borders', which resulted from his experience in Lille: did he not ascertain that the north-eastern borders of France were perfectly 'artificial'? Starting in 1948, Dion occupied the Chair of Historical Geography of France at the Collège de France. He preferred to call it 'retrospective human geography'. The main subjects of his lectures were the original sites of urban areas, the main trade routes in Gaul, the ancient role of river shipping, the exploitation of forests and, always, the origin of rural landscapes. In the 1960s, however, new themes appeared. Looking beyond Gaul, Dion wondered about the concept of geographical space in the Greco-Roman world. What was the extreme west for the Greeks? What was the 'geographical' value of Homer's poems? What was the meaning of Ulysses' voyages? Where exactly did Pythéas want to go? What were the geo-strategic concepts of Alexander or Caesar?

In a difficult, because subtly nuanced, work that is his scientific legacy, *Aspects politiques de la géographie antique* (*Political aspects of ancient geography*) (1977), Dion summarized his concepts and his hypotheses. His main idea is that ancient geographers and discoverers always served political authorities. Two examples are provided. First, the massaliot navigator Pythéas did not venture into the northern Atlantic all the way to Ultima Thule (Iceland or the Shetland Islands) out of mere curiosity. He had certainly been asked by the Macedonian government leaders to reconnoitre the extreme western limits of the ecumene. Did not Alexander the Great, after having conquered Asia, dream of conquering Europe all the way to the western ocean? Second, Caesar and his successors used geographical disinformation as a governing tool. Why would Caesar represent the Atlantic coasts of Gaul as facing north instead of west? According to Dion, Caesar needed to convince the Roman Senate that it was the best base from which to prepare a military invasion of (Great) Britain. Shortly thereafter, the geographer Strabo used this erroneous depiction of Gaul to avoid contradicting the 'divine Caesar'.

3. Influence and Spread of Ideas

Dion's influence over French geography has been considerable but it has been episodic. One can assert that until 1960 every French geographer had read *Le Val de Loire* (*The Loire valley*) and *L'Essai sur la formation du paysage rural français* (*Essay on the creation of French rural landscape*) and considered the two works as a supreme conclusion to Vidal de la Blache's thoughts. After 1960, Dion and geographers seem to part ways for various reasons. Dion never wrote any student textbooks or manuals. Teaching at the Collège de France distanced him from the universities where young researchers prepared their theses. Rural geography, after 1960, emphasized the study of economic and social issues in agriculture over rural landscapes. Finally, most French geographers no longer felt a concern for Dion's purely historic works. After 1980, they were, one could say, rejuvenated. His main publications were reprinted: *L'Histoire de la vigne et du vin en France* (*The history of vines and wine in France*) (1977), *Le Val de Loire* (*The Loire valley*) (1978), *Les Frontières de la France* (*France's borders*) (1979), a series of works on *Le Paysage et la vigne* (*Landscape and vines*) (1990), and *L'Essai sur la formation du paysage rural français* (*Essay on the creation of French rural landscape*) (1991). Mostly, Dion's work has initiated a renaissance of historical geography in France, a long-neglected speciality, as witnessed by *L'Histoire du paysage français* (*History of the French landscape*) by Jean Robert Pitte (1983) or *Géographie historique de la France* (*Historical geography of France*) by Xavier de Planhol (1988). Dion's influence has also been felt by historians of ancient geography. Raymond Chevallier paid him a just tribute in 'Mélanges offerts à Roger Dion' (1974). Claude Nicolet wrote in *L'Inventaire du monde: Géographie et politique aux origines de l'empire romain* (*Inventory of the world: geography and politics in the origins of the Roman Empire*) (1988): 'My debt to R. Dion's works on Ancient Geography, in particular that covering the reign of Augustus, is evident'. Younger authors, however, underline the dubiousness of certain hypotheses: 'analyses and methodology to be followed with caution' wrote Christian Jacob in *Géographie et ethnographie en Grèce antique* (*Geography and ethnography in Ancient Greece*) (1991).

Roger Dion was a discreet and somewhat marginal research worker, who had difficultly finding his niche between history and geography. His work is, however, testimony to the extremely close association of these two disciplines in the French tradition.

Bibliography and Sources

1. REFERENCES ON ROGER DION

Obituaries

Le Roy Ladurie, E., 'Roger Dion', *Annuaire du Collège de France*, 1981–1982, 71–2.

Broc, N., 'Roger Dion, 1896–1981', *Annales de Géographie*, xci (1982), 205–17.

Gourou, P., 'Roger Dion, 1896–1981', *Journal of Historical Geography*, 8, 2 (1982), 182–4.

Other Studies

Chevallier, R., 'Mélanges offerts à Roger Dion', *Littérature gréco-romaine et géographie historique*, Picard, Paris, 1974.

Bonnamour, J., 'Roger Dion', *Deux Siècles de géographie française* (1984), 174–9.

Jeanneau, J., Le Val de Loire des géographes. Roger Dion', *Actes du colloque d'Angers, Loire-Littérature*, mai 1988 (published 1989), 593–601.

Pitte, J.-R. 'Un géographe du vouloir humain', in *Le Paysage et la vigne* (1990), 7–20.

Le Paysage et la vigne. Essais de géographie historique, Editions Payot, Paris (1990).

Roncayolo, M., Une Leçon de géographie', in *Le Paysage et la vigne* (1990), 271–94.

Flatrès, P., 'Roger Dion: Une carrière, une oeuvre', in *Essai sur la formation du paysage rural français*, Flammarion, Paris (1991), I–XX.

2. SELECTED WORKS BY ROGER DION

Books

1933 *Le Val de Loire*, Arrault, Tours, 1933 (new edition Laffitte Reprints, Marseille 1978).

1934 *Essai sur la formation du paysage rural français*, Arrault, Tours, 1934 (new edition 1991).

1947 *Les Frontières de la France*, Hachette, Paris, 1947 (new edition 1979).

1949 *Paris dans les récits historiques et légendaires du IXe au XIIe siècle* Arrault, Tours, 1949.

1952 *La Création du vignoble bordelais*, Editions de l'ouest, Angers, 1952.

1959 *Histoire de la vigne et du vin en France des origines au XIXème siècle*, Ed. R. Dion, Paris, 1959 (new edition 1977).

1961 *Histoire des levées de la Loire*, Ed. R. Dion, Paris, 1961.

1969 *Les Anthrophages de l'Odyssée*, Editions Vrin, Paris, 1969.

1977 *Aspects politiques de la géographie antique*, Les Belles-Lettres, Paris, 1977.

Articles

1938 'Orléans et l'ancienne navigation de la Loire', *Annales de Géographie*, xlvii, 128–54.

1939 'Les Principaux types du paysage rural', in *La Campagne*, under the direction of R. Blais, 46–81.

1943 'Grands traits d'une géographie viticole de la France', *Publications de la société de géographie de Lille*, 5–69; 1948–9, 6–45.

1946 'La Part de la géographie et celle de l'histoire dans l'explication de l'habitat rural du bassin parisien', *Publications de la société de géographie de Lille*, 5–80.

1948 'Leçon d'ouverture du cours de géographie historique de la France prononcée au Collège de France le 4 décembre 1948', *Annuaire du Collège de France*, 1948–1949, 1–19.

1950 'Les Origines du vignoble bourguignon', *Annales (Economies, Sociétés, Civilisations)*, 433–9.

1951 'Réflexions de méthode à propos de *La Grande Limagne* de Max Derruau', *Annales de Géographie*, lx, 25–33.

1953 'Le problème des Cassitérides', *Mélanges Philippe Arbos*, 23–9.

1961 'Le "Bon" et "Beau" Pays nommé Champagne pouilleuse', *L'Information Géographique*, 209–14.

1966 'Où Pythéas voulait-il aller?' *Mélanges André Piganiol*, 1315–36.

Numa Broc is Professor of Geography at the University of Perpignan. (Translation of this biobibliography was rendered by Anne Marie d'Hauteserre, currently with the Department of Geography, Southern Connecticut State University.)

Chronology

1896	Born in Argenton-sur-Creuse
1902–13	Secondary studies in the Lycée de Blois
1913–15	Studies in the Lycée Louis-le-Grand, Paris
1916–18	Soldier in the First World War; awarded the *croix de guerre*.
1919–22	Student at the Ecole normale supérieure
1921	Aggregation – history and geography
1922–4	Member of the Thiers Foundation, Paris
1933	Doctor of Letters, Paris. Defended thesis *Le Val de Loire*
1934–45	Professor, University of Lille
1945–8	Professor, The Sorbonne
1947	Professor at the University of São Paulo; published *Les Frontières de la France*
1948	Professor of the Historical Geography of France at the Collège de France
1968	Retired
1981	Died in Paris

Ferdinand Magellan (Fernao de Magalhaes)

c.1480–1521

Sarah Lumley

Reproduced with kind permission of the Museo Naval, Madrid.

His name is written forever in the geography of the earth [the Straits of Magellan] *and in the heavens above* [Magellanic Clouds].
(Guillemard, Francis, *Life of Magellan*. Hakluyt Society, London, 1890)

Magellan is in many ways an enigma. Little is known about the details of his early life, and even the year and place of his birth in Portugal are not recorded with certainty. Contemporary accounts of the man attribute conflicting personality traits to him. After his death he was described by a supporter as noble, brave, unselfish, humble, democratic and pious, but his behaviour in life also showed him to be rash, arrogant, egotistical, domineering and lacking in judgement at crucial moments. This lack of judgement was to lead to his early death. Magellan made a significant contribution to geography and geographical thought by expanding European knowledge of the world that lay beyond the boundaries of Europe. He is widely believed to have named the Pacific Ocean, which he traversed in 1520–1 and which, until his voyage, was largely unexplored. It had been known, until then, as the South Sea (a name which was retained in common usage for nearly 200 years). Magellan as geographer enabled the borders of newly discovered countries and an ocean to be drawn upon the global map. He discovered for Europe important phenomena which now bear his name. He claimed for Spain that archipelago which came to be known as the Philippines, thus altering the religious, cultural and political landscape of Asia by establishing the foundation of the only predominantly Roman Catholic nation in the continent. For many years the Philippines was a protectorate of Mexico and as such has links with the history, culture and economy of Spain's former Central and South American colonies. To this day Magellan's memory is loathed by some and admired by others. His courage in life and in death is unquestionable, as is his hubris. Some scholars believed he was not accorded the recognition he deserved because, they argued, he was regarded as a traitor by Portugal and as Portuguese by Spain. Some even considered him to be a double agent. Insidiously, Magellan's voyage, and his discovery and claiming of what was to become the Philippines for Spain, made many contributions to contemporary geography and geographical thought. Magellan's contributions are seen as both positive and negative, witting and unwitting, laudable and contemptible. They are contributions to cartography and to physical, economic, political, historical and environmental geography. They may be viewed as the contributions of a hero and a villain.

1. Education, Life and Work

INTRODUCTION

Ferdinand Magellan (Fernao de Magalhaes) was born around 1480 and his birthplace is variously recorded as Oporto, Sabrosa or Ponte da Barca, in Portugal. The landscape of Magellan's childhood has been describe both as a barren countryside in northern Portugal and as natural surroundings more lovely than those he ever gazed upon in his long wanderings. It is difficult to find consistent descriptions of Magellan's life: of his birthplace, of his antecedents, of his descendants, of his personality or of his loyalties. Much of what has been written is conjecture, speculation or even fiction. Records and reports of his background, motives and achievements, whether they be contemporary with Magellan or later, seem to be coloured by the political or religious perspective of the writer concerned.

There is no doubt that Magellan's family was minor nobility and his paternal grandmother was probably from the important de Sousa family. In addition, Magellan was descended from a famous eleventh-century Crusader who bore the de Magalhaes name. Magellan is believed to have been the youngest of three children (his elder brother was probably known as Diogo de Sousa, and his sister is referred to sometimes as Isabel and sometimes as Teresa). In 1517 Magellan married Beatriz (Brites) Barbosa, who was from a wealthy and powerful family, and they probably had a son Rodrigo, who had no offspring, and a daughter named Anna de Magalhaes, from whom his line continued. Magellan lived in times which were rich in expedience, intrigue and power struggles. The Catholic Church was of great significance in matters of both politics and religion, and it was virtually impossible to separate the activities of the Church from those of the state. The Church had an overwhelming influence on day-to-day life in the Catholic nations of Europe, and on the life, achievements and errors of Magellan. Magellan was a man of the Renaissance but not, in modern terms, a Renaissance man. He was single-minded about his goals in life, and he chose the path that would lead him most readily to what he wanted to achieve.

GEOPOLITICAL BACKGROUND TO MAGELLAN'S TIME AND PLACE

At the time of Magellan's birth, the various nations of Europe had been experiencing feuds, invasions and civil insurrections for centuries. Until 1474, what became Spain had been under the separate Crowns of Aragon and Castile. Castile in particular had a long and bitter enmity with Portugal which, despite the best intentions of dynastic marriages and treaties, had never really been resolved and which 'set the geopolitical pattern in the earlier Iberian phase of Pacific history' (Spate, 1979, p. 26).

For nearly a century to Magellan's birth, the Muslim Ottoman Turks had been aggressively expanding their empire into the Balkans and Greece, and, briefly, into Italy (1480–1). The Moors had ruled Portugal for six hundred years, leaving only when Henry the Navigator was a boy in the early fifteenth century, and King John I of Portugal carried his war across to Morocco, the Moors' homeland. The Christian nations had been fighting with each other: England, France, Portugal, Castile, Aragon, Poland, Bohemia and Hungary had all been involved in war, often in civil war. Portugal had experienced peace within its own borders from 1385 to 1485 and in this it was unusual. A profound animosity between Christianity and Islam had been simmering since the Crusades began, but the Crusades provided Europe with an awareness of the wealth and mystery of Asia. Mercantilism was growing out of feudal Europe, marine technology was improving and Asia and Arabia held a monopoly over spices for which the European courts were growing greedy. The papacy was extremely powerful and much of this power could be used by whichever Catholic monarch had the Pope's favour. Only the Pope could endorse missions to heathen countries and such authority was vested in specific rulers. Spate (1979, p. 26) states:

> the Bull *Pontifex Romanus* of 1455 was accorded to Prince Henry [the Navigator, of Portugal] in his capacity as Governor of the Order of Christ [awarded to him for keeping the Moors in check], itself a survival from the Reconquista of the Peninsula. The salvation of unbelievers obviously might depend on secular strength, and that in turn on economic resources; mission rights, at least in the view of their recipients, carried with them as a necessary corollary rights of exploitation, and these could be best secured, perhaps only secured, by a monopoly in favour of the power behind the mission.

Essentially, the Pope perceived himself as having the power to grant world exploration rights, and with these the rights to the conversion of souls to Christianity which was inextricably linked with the exploitation of all those souls' resources (and particularly their wealth).

Rodrigo Borgia (of the infamous Borgia family) became Pope Alexander VI in 1492. Pope Alexander needed the help of Ferdinand and Isabella of Spain to promote his son Cesare to a principality in Italy. The Pope was completely malleable as far as Ferdinand and Isabella were concerned and was prepared to modify Papal Bulls in order to please them. Existing Bulls regarding imperial expansionism tended to be in Portugal's favour, but from May to September 1493 the Spanish rulers managed to obtain a series of papal Bulls which nearly nullified Portugal's claims. The ultimate Bull in this series was *Dudum siquidem*, which granted rights to Spain over a comprehensive part of the globe. Portugal was left with some forts in Africa and some Atlantic islands but its claims to exclusive rights in the Indies were effectively nullified. The zones drawn across the Atlantic Ocean meant that Spain could attempt to go west into the Orient but all other European countries were excluded from such exploration by what was, effectively, an international law. The King of Portugal tended to ignore the new Papal bulls. Characteristic of all the rulers of the day, he had spies in high places in the courts of his opponents. In this case he had knowledgeable and clever diplomats in the Spanish court, and he was able to negotiate a treaty direct with Spain, using inside information. The Treaty of Tordesillas, signed in 1494, placed the 'demarcation line at a position 370 leagues west of the Cape Verdes; and Alexander's jurisdiction was specifically set aside'. This demarcation was to lie at the heart of Magellan's later actions:

> since the whole Luso-Castilian concept of zones was predicated on eastwards and westwards voyaging to the Indies and Cathay, and obviously these voyagings could converge, the presumption grew up that the division must apply on the other side of the globe ... and when both Spaniards and Portuguese should reach the Moluccas [the Spice Islands] the twain would have met and the question become acute. (Spate, 1979, p. 26)

The question would become acute because with this expansion in opposite directions came the realization that the very location in which the twain would meet was probably the fabled Spice Islands, the land both parties most desired. At this stage the problem of fixing the longitude of the antipodal demarcation line became extremely urgent, but no-one at that time knew how to fix longitude reliably.

The promise of wealth and fame, colonial expansion, the desire for knowledge and the intense competition between Spain and Portugal which motivated many great advances in cartography, cosmography and geography took place in the generation before Magellan. By Magellan's time, and with much the same motives, a great goal of navigators was to circumnavigate the earth, to find new passages to distant lands, to expand the map of the known world and to demonstrate to the great thinkers and rulers of Europe, once and for all, that the earth was, in fact, round.

THE FORMATIVE YEARS AND THE VOYAGE

Much has been written, from many perspectives, of Magellan's life. There is little point in reproducing it here, save for the points which are salient to this reinterpretation of his contribution to geography, particularly to the geography of Asia. There follows a brief précis of his life, and of the incidents that led to his death (perhaps symbolically) at the hands of Chief Lapu-Lapu of Mactan off the coast of Cebu, in what was to become the Philippines.

At the age of about twelve years, Magellan went into service as a page to the household of Queen Leonor in Lisbon, where he was educated and trained. Perhaps significantly, at the time when Magellan entered the beginning of what was to be his apprenticeship in navigation (1492), Columbus was making his voyage of discovery to the Americas. Here he learned cartography, astronomy and navigation under the King's brother-in-law (brother of the Queen) Duke Manuel, son of Ferdinand and Isabella of Spain. In 1495 King John was poisoned and Duke Manuel became king. (This was not of advantage to Magellan, who was identified with the dead king's faction.) In 1496 Vasco da Gama was appointed leader of an expedition to India whence he returned with a cargo of spices, having found the sea route to India. Magellan was then employed as a clerk in the maritime division of what later became India House. In March 1505 Magellan sailed from Lisbon with Dom Francisco de Almeida's fleet, en route to India. He was classed as a 'gentleman adventurer of the King's household'. The fleet, led by the ship *San Miguel*, went via the Mozambique Channel off the East African coast. Part of the plan was to drive the Arabs out of the Indian Ocean so that all European trade to and from India could be monopolized by Portugal via the Atlantic Ocean to Lisbon. This plan eventually succeeded.

Magellan arrived in India in 1507. In September 1509, ships from the fleet visited Malacca. The Portuguese perceived that the spice trade was in the hands of Arab monopolists and that Muslim influence was very strong on the Malay peninsula, given the number and size of mosques visible in the area. The animosity between the Christian Portuguese and the Muslims (such as the Arabs and the Malays) was still simmering, and fighting broke out, with many men from the fleet being captured or killed in Malacca.

In January 1510 Governor Albuquerque commanded a Portuguese attack on Goa. The Portuguese were defeated, Albuquerque was nearly killed and Magellan was badly wounded. In 1513 Magellan participated in the capture of Azamor in Morocco, where he was wounded in the leg; he limped thereafter. Subsequently, Magellan asked King Manuel for an increase in pay and recognition of his services in the Indies and Morocco. Magellan was no favourite of Manuel's. The two had an historical dislike of each other following Magellan's page days and the death of the previous king, and King Manuel refused these requests, insulting Magellan's pride. Some writers argue that this disagreement eventually caused Magellan to offer his services to Castile, others that the move was motivated by Magellan's ambition to find a westward route to the Indies (and the desirable but elusive Spice Islands), which he could not do for Portugal. Another route to the Indies would have been in direct competition with Portugal's now firmly held African route and in violation of the Treaty of Tordesillas. By this argument, if Magellan were to discover such a route west he would have to do it for Spain. Magellan's true motive for his change of allegiance may well have been a combination of both factors.

There are also mixed views about how severely Portugal would have viewed Magellan's defection, and also how severely he has been judged by history. Columbus, Vespucci, Cabot and others also entered foreign service, and in those days it was not uncommon for allegiances to be transferred. Fleet personnel were often constituted of men from many nations, as was the crew of Magellan's own fleet. However, one writer (Zweig, 1938), claims that Magellan was a traitor to his country, a nobleman who 'was unpatriotic when he smuggled across the frontier naval secrets which he had learned through being given access to Tesorania in Lisbon'. All this notwithstanding, Magellan formally became a naturalized Spanish citizen in 1517. In this he was joined by Ruy Faleiro, a talented cosmographer. Between them they had decided that the Spice Islands were on the Spanish side of the Tordesillas line on the other side of the world. (Faleiro was later to develop a reputation for genius and mental instability.) Meanwhile, Magellan sought support for his life work, and in this he was successful. Juan de Aranda, an official of the Casa de Contracción in Seville, took seriously Magellan's claim that he could voyage westwards to the Spice Islands, and he gained the ear of the head of the Casa, which organized the Antillean Indies from Seville, the powerful bishop of Burgos, Juan de Fonseca. The entire route of the planned voyage south-west is not known but one of Magellan's chroniclers who accompanied him on the voyage, the Venetian, Pigafetta, claimed that Magellan had decided on a circumnavigation (straying through the Portuguese zone, which seems dangerous for a Portuguese defector). If Magellan were to achieve the goal of finding a westward route to the Spice Islands, he and his supporters would have the chance to become rich, especially with mission rights giving access to the resources of the souls who were converted to the Roman Catholic faith. Magellan is reputed to have been a pious man, so he no doubt believed in this type of missionary work. But given his naval history in Africa and the Indies, fighting Muslims, the potential to spread his own religion through territories in which Islam was moving must have seemed appealing.

Magellan eventually obtained the support and the ships he required, and after delays and obstacles had been surmounted, and the run-down ships had been refitted, Magellan's fleet left Seville in August 1519. There were five ships in the fleet but only two were to return to Spain, only one, the *Victoria*, having completed the voyage. Magellan's flagship was the *Trinidad*, the other ships in the fleet being the *San Antonio*, the *Victoria*, the *Concepción* and the *Santiago*. The *Santiago* was lost in July 1520, and the *San Antonio* deserted the fleet in November 1520, returning to Seville in May 1521. The *Concepción* was burned off the Philippines in April 1521, and the *Trinidad* was deemed unseaworthy in December 1521. Alone, the *Victoria* struggled back to Seville in 1522, having circumnavigated the world. She had lost her captain-general, all of her sister ships and a large proportion of her crew, but with her return the geography of the world known to Europe was changed forever.

From early in the journey there was friction among the multinational crew and a prevailing suspicion of the formerly Portuguese Magellan, who seemed to be taking the Spanish fleet dangerously close to Portuguese waters. (Indeed, Gaspar Correa's contemporary writing demonstrates the animosity of the Portuguese towards Magellan.) Juan de Cartagena, the captain of the *San Antonio*, was removed from his post for insubordination not long after the fleet moved below the equator, close to the African coast. De Cartagena was placed under arrest. In January 1521, having followed a southerly course to Brazil, the fleet reached the la Plata estuary between Argentina and Uruguay, but found no passage through the South American continent there. They sailed on down the South Atlantic along the east coast of the continent. At San Julián on the Patagonian coast of Argentina, where Magellan had decided to overwinter, the officers mutinied. On the day of the Easter Mass on the flagship, de Cartagena and de Queseda seized the *San Antonio* (captained by Magellan's cousin Álvaro de Mesquita, who was attending the mass), the *Concepción* and the *Victoria*. When the mutiny was quelled, forty men were sentenced to death, but finally only Quesada was executed (and his body mutilated in a horrible manner), while Cartagena was sentenced to be marooned on the Patagonian coast with a priest complicit in the mutiny, Fr Sanchez de Reina. The other men were pardoned, mainly because, making up one-sixth of the fleet's crew, managing the ships without them would have been impossible. One of the pardoned men was Juan Sebastián del Cano, who attained fame after the last leg of the voyage when he completed the circumnavigation. Generally, history has not laid the blame for the mutiny at Magellan's feet. He is perceived to have been an overbearing and authoritarian man, but as captain he was supposed to give orders and maintain a strong control over the fleet. At this time everyone on the ships was suffering considerable hardship through cold and lack of food, and Magellan is deemed to have been reasonable in his behaviour under the circumstances.

Magellan was anxious to leave the coast of Patagonia because he did not believe they could survive the long winter there, and in July 1520 he sent Juan Serrano out in the *Santiago* to try a southern run. The ship ran into a storm, lost its rudder and foundered. Bar one, the entire crew survived but the ship was lost. Conditions for the survivors were appalling but two men reached Magellan by an overland route, walking through dangerous territory for eleven days, and alerted rescuers to the fate of the ship. The chronicler Pigafetta was with those who took supplies to the survivors, and he provided detailed descriptions of the cold and the difficult terrain in which they were stranded.

The fleet left San Julián in August 1520 and soon ran into another storm, from which it took shelter at Santa Cruz, whence it departed in October. Passing the Cape of Eleven Thousand Virgins (named by the pious Magellan in recognition of a Saint's day), they came to a bay which Magellan hoped might offer a passage through the continent. He ordered a thorough investigation of what appeared to be a mountain-ringed dead-end, sending *San Antonio* and *Concepción* to explore. To the north of Tierra del Fuego, the ships found a passage, and the Strait of Magellan (Estrecho de Magallanes) was discovered. The continued search for this passage and its discovery is indicative of Magellan's tenacity, thoroughness and sense of certainty; in this instance he was also lucky. Later, all of these characteristics, except luck, were to lead him to death. While reports of the details of Magellan's life are often inconsistent, observations about his personality generally agree. Clearly, he had a strong personality. He had a great will to succeed in all his ventures, and he was no democrat. He imposed his will on others ruthlessly. It seems that his belief in his own religion was unassailable, as was his belief in the existence of the passage through South America. He believed just as firmly in the Spice Islands and in his own notion of how to reach them. He seems to have placed the realization of his personal goals above his own life and that of others. It would not be unfair to say that he was hot-tempered, stubborn, ruthless and reckless. His judgement of the layout of the physical world and his understanding of navigation was impeccable. His judgement of how to behave to his fellow men, whether those of his own race or those of other races and cultures, at least to a modern view, was somewhat lacking.

After the elation of discovering the Strait of Magellan, there was dissension about whether to turn back to Seville or to continue in the pursuit of the Spice Islands. As far as Magellan was concerned, the position was not negotiable: he was continuing and so was his fleet. However, there was a very long sea journey ahead and only three months' supply of food remained. Fleet members had already endured severe hardship and many wanted to return home. Beyond the Strait of Magellan lay the unexplored and un-named Pacific. It was the largest stretch of ocean on earth. Near the mouth of the Strait of Magellan, the *Trinidad* and the *Victoria* waited six days for the *Concepción* and the *San Antonio*. They searched for and found the *Concepción*, but after fruitless searching realized that the *San Antonio* had deserted: the largest ship in the fleet had turned its back on them, taking with it the greatest portion of the already meagre food supplies. On 28 November 1520 the three remaining ships sailed into the Pacific. The *San Antonio* sailed on to Seville, arriving in May 1521 with outrageous and false tales of Magellan's actions. From the deserters' reports, Magellan was identified as a political risk to Spain and his family was placed under observation to prevent them running to Portugal. Since Magellan had sailed though Portuguese territory to reach the Strait of Magellan, the deserters' tales were plausible and King Charles of Spain did not claim the right to use the route.

Meanwhile, Magellan and his ships had sailed on into the Pacific, which Magellan is believed to have named on account of the good weather he encountered there. The men were slowly starving over the nearly four months it took to reach Leyte in the Philippines.

Magellan seems to have steered his ships across the Pacific towards the Malay peninsula in order to take bearings in a familiar area before continuing his search for the Spice Islands. By the time he reached Guam the men were desperate for fresh food and water, and although they managed to get some supplies, unwonted aggressive behaviour by Magellan and his men towards the Guamanians and a desire to reach the South China Sea meant that the fleet left the area abruptly. Reports of this incident show Magellan in a bad light. On 16 March 1521 Magellan reached Leyte Gulf in the eastern Visayas. He put to shore on the isolated island of Homonhon, where those sick or dying of scurvy were nursed. Fresh water and coconuts succoured the sickly crew. While they were on Homonhon, friendly islanders visited in boats and Pigafetta wrote of his favourable impressions of them. These people fetched food from Leyte for the ship's men. Later, Magellan celebrated an Easter Mass on Limasawa – the first mass offered on Filipino soil – with some of the bewildered local people present. Here Magellan erected a large wooden cross which, some say, served as a way of claiming the territory for Spain.

Magellan still had a pressing need for food and was advised to go to Cebu. On 7 April 1521 the ships entered Cebu's port. Cebu was a large, sophisticated and wealthy town and Magellan foolishly decided to enter the port with ceremony, firing guns and waving banners, frightening the local people and appearing pugnacious. Magellan had a Malay interpreter with him and was able to converse with the chief of Cebu. However, the difference in language and culture must have been an insurmountable barrier to real understanding. Magellan's visit to Cebu culminated in him trying to implement a proposal to make all the people of the archipelago Christian subjects of Charles of Spain, with Chief Hamubon of Cebu being the overlord of all the other chiefs in the area (a concept which was completely alien to the people, over whom the chiefs had limited authority anyway). There were celebrations and mass baptisms. The islanders were commanded to abandon their religion and forfeit their idols. It is hard to believe that the chief or any of his people were serious about submitting to Magellan; perhaps they were courteously enjoying the celebrations. Magellan appears to have misinterpreted the situation and, as Noone says:

> The smouldering resentment at the Spaniards' intrusion into the peoples' lives, the unwanted pressure regarding baptism, the anger of the chiefs compelled to submit to the novelty of a paramount chief would build up and violently explode before many days were out. (Noone, 1982, p. 80)

Also, according to Pigafetta, 'some of the chiefs refused to submit to Hamubon or to us' (Noone, 1982, p. 81). One of these chiefs was Lapu-Lapu from neighbouring Mactan, one of two chiefs from Mactan. The other Mactan chief, Sula, submitted and asked Magellan to attack Lapu-Lapu with him the next night. Magellan, his authority challenged and his pride stung, agreed. In what is now a well-known tale, Magellan was killed in the battle (which again he misjudged) on 27 April 1521, having been hit in the leg with a poisoned arrow. Pigafetta, who was a loyal supporter of Magellan, left an account of his Captain-General's courage, which describes how Magellan sacrificed his own life to save his men. Since Pigafetta's chronicle was not translated into Portuguese until 1938 (or published in Pigafetta's native Italian until long after Magellan's death), Pigafetta's praise and defence of Magellan were not widely known among his detractors. Thus Magellan, after his death, continued to be viewed as a hero or villain, brilliant explorer or madman, treacherous to Portugal or Spain, noted natural son of one and adopted son of the other. However he is viewed, and whatever his contribution to the world, there seems to be no doubt that he had courage. It seems appropriate to end this summary of his life with Pigafetta's words on his death, which

> killed our mirror, our light, our comfort and our true guide. While they were wounding him, he turned round many times to see if we all had reached the boats. Realising he was as well as dead, and wounded as we were, we retreated to our boats which were already pulling off. (Noone, 1982, p. 84)

After more treachery and mishaps, the only ship remaining, the *Victoria*, reached Seville, via the Spice Islands, in September 1522. Here, for various political reasons, the tarnishing of Magellan's reputation continued, though that of del Cano shone. The Spaniards, though not returning for a while, did not forget their new colony. In August 1543 another expedition was sent to the islands and the Philippines were named in honour of King Charles' son Philip. In 1564 the Basque navigator Miguel Lopez de Legaspi led a third expedition to the Philippines, this time with the express purpose of fully conquering and Christianizing the islands. Again there was resistance, but this time, after a struggle, the conquest was deemed to be complete.

2. Scientific Ideas and Geographical Thought

Because of Magellan's early and sudden death, he had no opportunity to write of his thoughts and ideas. All that remain of his own writings are his wills and a few letters and papers (some of suspect authorship). Perhaps he would never have summarized his thoughts and achievements had he survived. However, although his life and death were beset by controversy, the impression of a man convinced of the correctness of his ideas prevails. Because of the political and religious power struggles of his own times, the significance of his achievements for science and geography were not always fully acknowledged. Whatever the prevailing wisdom of the day, and the retrospective interpretations of his motives by historians, it seems that the strongest reasons for his actions were to test his hypotheses about the passage through South America, and to locate the position of the Spice Islands (Moluccas). His hypothesis about the passage through South America was proved correct, and he came very close to finding the Spice Islands, and almost certainly would have done so had he survived (as indeed did his remaining crew, following Magellan's ideas). Magellan's transgressions of Portuguese zones and his apparent intention to circumnavigate the globe regardless of demarcations seem to have been motivated, more than by other influences, by his desire to test his hypotheses. Magellan was without doubt an excellent navigator and his insights seem to have had a sound basis in science. He had on his ships, mutineers and deserters aside, men of scientific distinction. While crossing the southern Atlantic, Magellanic Clouds (named after the captain-general) were discovered. These are 'two small satellite galaxies of the Milky Way which are visible in southern circumpolar skies, where they appear as cloudy stellar concentrations'. In addition, the long-sought passage through South America bears the name the Strait of Magellan, and though not exploited directly by Spain, became important for the Dutch in their own pursuit of the spice trade. His scientific insights and stubborn persistence brought him to within a few hundred kilometres of the Moluccas (Maluku), which lie south of the Philippines in Indonesia. The islands were colonized by the Portuguese soon after Magellan's voyage but were taken by the Dutch in 1605, and it was the Dutch who subsequently monopolized the spice trade. Without Magellan's voyage the chronicles of his fellow travellers, which contributed to geographical knowledge, would not have been written. In particular, the detailed and observant (though sometimes biased by loyalty to Magellan) chronicles of Pigafetta recorded the physical and human aspects of the world through which he travelled. He wrote of the subsequently lost religion and culture of the Philippines, without which our contemporary knowledge of such matters would be much poorer. By scientific calculation, and by intuition about the size and shape of the earth and the relationship between land masses, Magellan also contributed to geographical thought.

Magellan's obvious scientific and geographical achievements are widely recorded. The phenomena which bear his name – Magellanic Clouds and the Strait of Magellan – are well known. His transnavigation and probable naming of the Pacific Ocean, the discovery of the Spice Islands and the successful circumnavigation of the globe are all recognized as great achievements. However, at least in Asia, some of his achievements might be perceived to have a darker side. His claiming of the Philippines, and its subsequent conquest for Spain and the Roman Catholic Church, has had the most enduring influence. Magellan's scientific ideas and the manner in which he set out to prove his hypotheses were inextricably linked with the Christianization and colonization of the territories which led, for him, to the Spice Islands. This territory, an archipelago loosely knit by history, culture and ethnicity, was to become the Philippines: a nation in its own right, and the only Roman Catholic nation in Asia with its closest neighbours being Muslims, Magellan's old adversaries. Magellan's heavy-handed attempt at mass baptism and his aim of establishing a paramount chief, answerable to the king of Spain, had its corollary. These efforts were to be subsequently translated into Christian conversion of the majority of the population, alienation of its Muslim minority, establishment of a powerful oligarchy, establishment of inequitable land tenure and four hundred years of colonial rule. By his death at the hands of the Mactan chief Lapu-Lapu, Magellan also provided a hero for a fledgling nationalist movement centuries later.

3. Influence and Spread of Ideas

Magellan's arrival in Leyte was to alter the geography of Asia and to start the slow physical and cultural decline of an archipelago which was rich in renewable resources and cultural diversity.

The archipelago now known as the Philippines is made up of 7100 islands spread over 1.29 million square kilometres of the Pacific Ocean. Ninety-six per cent of the land area is occupied by 11 of these islands. This oceanic archipelago had been visited and settled by seafaring people for thousands of years. Topography, rainfall and soil type in the islands vary and vegetation ranges from tropical rainforest to grasslands. There are extensive river systems in some of the islands, such as the Pulangi and the Agusan in Mindanao and the Cagayan and Pampanga in Luzon. In Mindanao, Surigao, Luzon and Cebu, copper, gold, zinc, nickel, iron, manganese and chromite are found. The distance from the northern tip of Luzon to the most southern point of the Sulu archipelago is 1625 km. To the west lies Malaysia (and Malacca, scene of some of Magellan's early adventures), to the south is Indonesia (and the Spice Islands), Japan lies due north and China lies north-west. People from all of these countries were familiar with the islands by the time Magellan arrived. The islands are generally mountainous and in Magellan's time they were thickly forested, with human settlement sparsely scattered around the narrow coastal lowlands, where the people lived in close association with the rivers, the sea and the forests. The original inhabitants of the Philippines whom the Spanish found on Negros and named the negritos (for obvious reasons) were ethnically separate from the more numerous later settlers from Malaysia and Indonesia, and are believed to have migrated over land bridges up to 45,000 years ago. Settlers from Malaysia and Indonesia came gradually over several thousand years and there is evidence of long trading relationships with the Chinese, Japanese and possibly the Arabs. About 130 years before Magellan arrived, Islam began to spread slowly through the islands, being introduced from what is now Indonesia. Magellan would have been aware of this and it may have explained his haste to baptise the people of Cebu. At Magellan's time the Sulu Archipelago and Mindanao were Muslim. The rest of the 500,000 inhabitants practised their own often sophisticated religions, some of which Pigafetta observed. When Magellan came, the lifestyle of the inhabitants was sustainable in the long term. Most of their protein was obtained from fish. They had a rich and diverse supply of food plants, some of which were native to the islands and some of which they brought from their countries of origin. Cultivation was usually conducted carefully in especially cleared areas of forest which, after harvest, were returned to the forest and not cultivated for many years.

The islands did not constitute a nation. Settlement was according to ethnic and cultural origins, with a hierarchical system based on kinship ties, where a datu or chief had a fairly loose authority over his settlement. (This was an arrangement which Magellan completely failed to comprehend when he tried to force one chief's authority over others.) There was no centralized ruler, and no system of centralized government. People lived in scattered and independent towns and villages. There were sometimes rivalries between different groups which the Spaniards used to their advantage in their conquest of the islands. Indeed, the lack of centralized rule was also used by the Spaniards, when they understood it, because it also meant there was no centralized resistance.

The people had a rich culture with a predominantly Malay and Indonesian influence. There was a high literacy rate, particularly among women, and the cultural heritage included stories, songs, fables and criminal and religious codes. Private property was unknown and land was defined in terms of how it was used rather than who owned it, since ownership was communal. Tenure was through usufruct, which meant that it was communally based, and that people had the right to use land but not to alienate it in any way. This was useful to the Spaniards later in their conquest (as was such tenure in other lands, to other European colonizers) because they took for the Crown all uncultivated lands, stating that, as they were unoccupied, they belonged to no one. In a parallel with their settlement practices in Central and South America, the Spaniards also later established the privately owned hacienda system and used it to create a landed oligarchy, while disenfranchising the rest of the population who had communal ownership claims to the lands.

Thus, when Magellan arrived and claimed the archipelago for Spain, he initiated a process which would inexorably change the culture and the physical environment of the Philippines over centuries, culminating in the nation as it is today: a country which has slid into an environmentally and socio-economically unsustainable condition. Thus alongside Magellan's creditable contributions to geography he also, by claiming the Philippines for Spain and uniting a loosely knit group of islands into one nation, made some less than desirable contributions to economic, political and environmental geography. Magellan's arrival heralded a change in geography, but it also signalled a change in thought. Attitudes to the land, politics and religion altered so that the new geographical thought came to encompass strictly hierarchical centralized power systems, alienation of commonly owned lands into massive private haciendas, and the advent of commerce on a global scale. Colonialism, rebellion, insurgency, population explosion, poverty, dictatorship and corruption, at least in the Philippines, resulted from this shift in geographical thinking.

Magellan's voyage opened a 'back route to Asia'. Various European nations had been exploring the region via the African route (recently secured by Portugal), but access to Asia both east and west from Europe began the process by which Spain, Portugal, Holland, France, Britain and, much later, the USA secured colonies in Asia. There is no doubt that Asian nations had had the tendency to invade and colonize each other for centuries, and that trade from China to Arabia had occurred over many years. However, Magellan's voyage heralded the beginning of a new era of colonization and exploitation in Asia, and indeed new ways of thinking about humans and their environment.

The Catholic Church and the Catholic population of the Philippines have no regrets about their religion, although some members of the Catholic Church appear to have misgivings about its historical missionary style. Noone (1982), writing for the Missionary Society of St Columban in a slightly apologetic way, states:

> Still another provision of [Pope] Alexander's decree, also to be put down to inexperience, was the bestowal of authority on the Spanish sovereigns (as had already been done in Portugal's case) to preach the Gospel, to send missionaries, to build churches, to do everything to spread the Faith among the people of the newly discovered countries – an arrangement that had mixed results, on the one hand magnificent in the Christianization of a large part of the Americas and the Philippines, as well as numerous enclaves of Christianity in Africa and Portuguese Asia: disastrous in identifying Christianity with 'colonialism' on the other. The most deplorable feature of all was that little account was taken of the feelings and rights of the newly discovered territories, it being considered axiomatic that Christian Europe knew what was good for them. (p. 14)

Without Magellan's voyage of discovery, the contemporary political, economic and environmental geography of Asia would have been different. The Philippines would probably have been Muslim, sustainable land tenure and property rights systems extant in Asia at that time might have prevailed, and the other European powers might have failed to establish their own Asian colonies. In addition, the Philippines' provision of a strategic and heroic base for the USA in the Second World War, and its placement of important US military installations in the Pacific, would not have existed without the USA's intervention in the country's war of independence from Spain in the late nineteenth century, after which America took control of the colony. We can only speculate about what could have happened. However, it is certain that many geographical ideas about Asia, particularly those concerning the Philippines, and the way in which the Asian countries are viewed by Europe, would have been different. Magellan's activities provided a precedent for a new political and geographical way of thought.

Whatever Magellan's virtues and failings, his name has become synonymous with epic voyages of discovery. He has caught the imagination of explorers in the age of technology. Magellan has a space probe named after him. It was sent out into the universe to explore the planets of our solar system. The Magellan Global Positioning System (GPS), a hand-held device which helps people on earth fix their location to within a few metres, is named after him, and armchair explorers can use the Magellan Internet Guide to navigate their way through the internet. Magellan's name has entered postmodern consciousness. It continues to be used to spread influence and ideas via a technology unimaginable to him, tossed as he was in his sixteenth-century ship, in the world's oceans, and in the intrigues, power games and bitter struggles which accompanied the human desire to conquer the spaces on earth.

Bibliography and Sources

1. SELECTED BIOGRAPHIES AND OTHER WORKS ON FERDINAND MAGELLAN

Noone, M.J., *The Islands Saw It. The Discovery and Conquest of the Philippines 1521–1581*, Missionary Society of St Columban, Helicon Press, Ireland, 1982.

Parr, C.M., *So Noble a Captain. The Life and Voyages of Ferdinand Magellan*, Robert Hale, London, 1955.

Pigafetta, A., *The Voyage of Magellan, the Journal of Pigafetta*, Translated by Paula Spulin Paige, Englewood Cliffs, New Jersey, 1969.

Pigafetta, A., *Magellan's Voyage: A Narrative Account of the First Navigation*, Translated and edited by R.A. Skelton, The Folio Society, London, 1975.

Spate, O.H.K., *The Pacific since Magellan*, Vol. 1, *The Spanish Lake*, Australian National University Press, Canberra, 1979.

Stanley, Lord of Alderley, *The First Voyage Round the World by Magellan. Translated from the Accounts of Pigafetta and Other Contemporary Writers*, The Hakluyt Society, Burt Franklin, New York, 1874.

Zweig, S., *Magellan Pioneer of the Pacific*, The University Press, Edinburgh, 1938.

2. SELECTED ARCHIVAL SOURCES

A number of sixteenth-century documents relate to the life and times of Magellan. Many of these have been copied and translated several times. The main contemporary accounts of Magellan come from those who travelled with him. The most significant narrative account is that of the Venetian, Antonio Pigafetta, whose writings are observant and detailed, and cover the years of the voyage. There are four manuscripts of Pigafetta's work. Three are in French and one is in Italian. Two manuscripts are in the Bibliothèque Impériale in Paris, one is in the Beinecke Rare Book and Manuscript Library at Yale University in the USA, and one is in the Biblioteca Ambrosiana in Milan. Pigafetta's petition to the Doge and Council of Venice, dated 5 August 1524, seeking permission to print his account of the voyage (which was refused), is in the Venice Archives. Early Portuguese versions of another well-known account, that of a Genoese pilot (whose name was possibly Bautista), are in the Bibliothèque Impériale in Paris, and in the Instituto de Historia y Cultura Naval in Madrid. A copy of a log book by Francisco Alvo is in the Manuscript Collection of the British Museum, while the original is in the Archivo General de Simancas in Valladolid. Magellan's 1519 will is in the Archivo General de Indias in Seville, while a work entitled *A Description of the Kingdoms ... from the Cape of Good Hope to China*, allegedly by Magellan (but of dubious authorship), is in the Madrid Archives.

Sarah Lumley has Honours and Masters degrees in Agricultural Science and is a lecturer in geography at the University of Western Australia. She previously lectured in geography at Deakin University in Victoria. Before joining academia she had an 11-year career in the Victorian Public Service, working in environmental and land use policy, and economics. She has a PhD (from La Trobe University) for her examination of the economic and historical reasons for unsustainable natural resource use in the Philippines.

Chronology

~1480	Magellan born in Portugal
1492	Becomes a page to Queen Leonor
1505–12	At war in the East Indies
1513	At war in Africa (at the capture of Azamor in Morocco)
1514	Quarrels with the Portuguese monarch
1517	Adopts Castile and renounces Portuguese nationality
1518	Signs an agreement with the King of Spain
1519	Magellan's fleet leaves for San Lucar (10 August)
1519	Fleet sails from San Lucar (20 September); reaches Rio de Janeiro (Brazil) (13 December)
1520	Arrives Rio de la Plata (Uruguay/Argentina) (10 January); reaches Port San Julián (Argentina–Patagonia) (31 March); mutiny in Port San Julián (2 April); loss of the *Santiago* (July)
1520	Departure from Santa Cruz (Patagonian Coast) (18 October); reaches entrance to the Strait of Magellan (21 October); crew of *San Antonio* desert with ship (November); fleet enters the Pacific (28 November)
1521	Reaches the Ladrones (Guam) (6 March); reaches Cebu (7 April)
1521	Magellan killed in Mactan (27 April)
1521	Survivors burn the *Concepción* (May)
1521	*Victoria* and *Trinidad* reach the Moluccas (8 November)
1521	*Trinidad* deemed unseaworthy (18 December)
1522	*Victoria* reaches Seville (8 September)

Kenneth J. Mason

1887–1976

Andrew S. Goudie

Kenneth Mason was a leading British explorer, surveyor and soldier who became the first holder of the Chair of Geography in the University of Oxford.

1. Education, Life and Work

Mason was born at Sutton in Surrey on 10 September 1887, the year of Queen Victoria's golden jubilee and of Mackinder's appointment to the Readership in Geography at Oxford. His father, Stanley Engledine Mason, was a timber merchant, and his mother, Ellen Martin, was the daughter of a City of London wine merchant. His first education was at Homefield Preparatory School, where he was savagely beaten. On one occasion he received six on each hand and an additional ten on his bottom. He regarded such treatment with equanimity, offering the opinion in his unpublished autobiography that 'If every silly ass that grows a beard and sits down in the London roads to demonstrate had been well and truly beaten when young, he too might have learned sense!' While at Homefield he had read such books as *Deeds that Won the Empire*, Younghusband's *Heart of a Continent* and Livingstone's *Darkest Africa*. These works helped to determine his career choice and so in January 1901, on the day that the Queen Empress died, he went to Cheltenham College, sometimes called the Gentleman's College to differentiate it from the even more famous Ladies' College. He did so because it had 'a greater number of old boys in the gunners and sappers than any other public school'. It was a public (i.e. private) school that prepared boys for the army examinations, and during his time a trio of Cheltonians also took the examinations: their names were Slaughter, Blood and Gore. Subsequently, two of these were seriously injured in action, and the other was killed. From Cheltenham, where mathematics was his best subject, he went in 1905 to the Royal Military Academy at Woolwich, from which he passed out in 1906 and was commissioned into the Royal Engineers as a second lieutenant.

The army sent him to the School of Military Engineering at Chatham, where he was instructed in such matters as surveying, mechanics, topographic sketching, musketry, revolver shooting, and ballistics. He also received instruction in stereo-photogrammetry. This training at Chatham fitted him for the rest of his army career. In 1909, having been promoted to lieutenant, he sailed for India, joining the Survey of India at Dehra Dun. In due course he was posted to survey Kashmir and the Western Himalayas, thereby gaining his love for the high mountains on the borders of the Indian Empire. He became a competent mountaineer, and he was, he believed, the first man to use skis on the Himalayas.

In 1913 he became involved with the Indo-Russian triangulation, the object of which was to join together the triangulations of India and Russia. He went to Gilgit and Hunza, travelling as fast as he could beyond the range of interfering telegraphs. He then went beyond Pasu, up to the Shimshal Gorge, and on to the Russian–China border and the Pamirs. His recreations on this trip included shooting animals, and his trophies are now on the walls of the Oxford School of Geography. They include the much sought after but greatly threatened Marco Polo sheep (*Ovis poli*).

In 1914, following the outbreak of the First World War, he was sent to the European front, where he witnessed the carnage at first hand. At Ypres in 1915 his small party of men, a dozen or so (Sikhs), was blown up by a high-explosive shell. He was the only survivor. He was sent to Osborne on the Isle of Wight (formerly Queen Victoria's home) to convalesce from his injuries, but returned to the trenches later in the year. Also in 1915, while on leave, he became engaged to Dorothy Robinson, whom he married in the garrison town of Aldershot in June 1917. In the meantime he had served in Mesopotamia, where he had a distinguished and eventful career. He was mentioned in despatches three times, and was awarded the Military Cross by George V for leading a night march to the flank of the Dujaihah redoubt. In the winter of 1917–18 he undertook reconnaissances in Kurdistan and a car crossing of the Syrian desert.

After the war he rejoined the Survey on India, but felt that he was under-employed. Thus while on leave in Britain in 1923–24 he talked to three major figures, Hinks, Younghusband and Longstaff, about an expedition to the Karakoram Mountains. In 1926, through their support, he was able to lead his Shaksgam Expedition. This was well received in the Royal Geographical Society, which awarded him its Cuthbert Peek award in 1926 and its Founder's Medal in 1927. In the same year he was appointed Assistant Surveyor General and posted to Calcutta. There he became a founder member of the Himalaya Club and the first editor of the *Himalayan Journal*.

Mason was by this time looking for a change of career, and while representing the Survey of India at the International Union of Geodesy in Stockholm he talked to Hinks, secretary of the Royal Geographical Society. In 1931, following on from that conversation, Hinks informed Mason, who was then in Burma, that Oxford University was to establish at long last a chair in geography, and suggested that he might apply. This Mason did, and used some distinguished individuals as his referees: Sir Sidney Burrard and Sir Gerald Lennox-Conyngham, both of whom had headed the Survey of India, Sir Aurel Stein, Sir Francis Younghusband, and Sir Charles Close. Oxford was impressed, and without being requested to attend for interview, he was duly elected to the chair from 1 May 1932. When he arrived in Oxford it was the first time that he had ever set foot in the precincts of a university. He became a Fellow of Hertford College and held the chair for 21 years until his retirement in 1953.

During the Second World War, Lieutenant-Colonel Mason, as he then was, became involved in intelligence work, as he had done in the later stages of the First

World War. In 1940, with Rear Admiral Godfrey and Ian Fleming (the creator of James Bond), he worked for the Inter-Service Topographical Department, which in 1941 became a part of the Admiralty's Naval Intelligence Division (NID). A large part of central Oxford, including the School of Geography, Manchester College, New College Library, the New Bodleian Library, and part of the Balliol College sports ground, became what was colloquially known as 'Admiralty Village'. The NID produced over 58 volumes of large and detailed regional handbooks, and undertook aerial photo analysis work for major campaigns such as the D-Day landings of 1944. At the end of hostilities he was offered an OBE for his services by the Labour prime minister, Clement Attlee, but declined it.

After his retirement, Mason undertook some travel, wrote his history of Himalayan exploration *Abode of Snow* (1955), and was involved with various institutions for which he had an abiding affection, including the Royal Geographical Society (but not the Institute of British Geographers), the Drapers Company and Cheltenham College. He died in Berkshire in June 1976 at the age of 88.

2. Scientific Ideas and Geographical Thought

SURVEY AND EXPLORATION

Mason's contribution to survey and exploration can be divided into two main components. The first was survey work in Kashmir, and the second was exploration in the Karakorams.

Mason's survey work in Kashmir was undertaken as an employee of the Survey of India and on the orders of Sir Sidney Burrard, who was Surveyor General from 1910 to 1919. As part of this work he climbed extensively. It was at this time that the government of India agreed with the Russian government to join up the triangulations of India and Russia, and this process was started in 1909. The work was to be of the very highest order of accuracy, and observations were made to heliographs placed on station mark-stones. After his predecessor Lieutenant H.G. Bell died of acute appendicitis, which struck at a height of 17,695 feet on Lup Gaz, Mason was put in charge of the work. He linked up with Russian stations on the Russian–Chinese boundary and worked southwards over the Taghdumbash Pamir. He fixed the position and heights of a number of peaks of the Western Karakoram. He also made extensive use of ground stereo-photogrammetry, a technique that he had learnt at Chatham. Mason's work undoubtedly facilitated subsequent exploration and mountaineering activity in this area.

In 1926 the government of India gave Mason permission to lead an expedition to explore the Shaksgam Valley, a great and remote tributary of the Yarkand that flows on the north side of the Great Karakoram Range and to the south of the Aghil Mountains. The expedition was a substantial one. Mason required 21 baggage ponies and 24 Ladaki porters for work in the Shaksgam itself for 82 days, but to march to the Shaksgam and to carry in supplies of equipment, food and fodder, he required a further 137 baggage ponies and 46 pony men to look after them. The party explored a large area of unexplored country, used stereoplotting techniques to make maps, and made observations of glacier-dammed lakes. This last facet of the work was something that Mason was to develop as part of his concern with hazards in these high mountainous environments.

HAZARDS AND APPLIED GEOMORPHOLOGY

During his traverses for the Survey of India, and as a result of his Shaksgam Expedition, Mason obtained a very detailed appreciation of the high mountain environments of Asia and used his surveying skills to set down the current state of some of the region's glaciers. He was intrigued by the many hazards that the area posed to human settlements, including the floods generated by the collapse of ice-dammed lakes.

This aspect of Mason's work has been reviewed by Hewitt (1989). He argues that Mason's role in monitoring of the Khumdan glacier dams on the Upper Shyok and of large outburst floods in 1926, 1929 and 1931 is of enduring interest, and points out that Mason was one of the earlier people to recognize the importance of 'accidental' glacier advances, which we would now call glacial surges. Mason did not believe that other sorts of glacier movement would cause great ice dams to occur, because such movements are slow, enabling rivers to keep a clear channel. However, he wrote that 'With "accidental" movements, the advance of the snout is often extremely rapid, so rapid that a large valley becomes completely blocked, a great lake several miles long is impounded, and a great flood may occur when the waters are let loose' (Mason, 1935, p. 25). In this same paper Mason gave a very useful review of the history of termini in the Karakorams, using the accounts of earlier glacier explorers and surveyors, and this was information that has proved useful to subsequent investigators such as the scientists of the International Karakoram Project of 1980 (see Goudie *et al.*, 1984). He pointed to the importance of topographic conditions in determining rates of advance, argued that 'a glacier which has once made a rapid advance is extremely likely to repeat the performance after some interval of time', and suggested various mechanisms that could cause sudden advances (e.g. augmented ice accumulation in the feeding area provided by ice-avalanches).

3. Influence and Spread of Ideas

Mason's greatest influence on academic geography was through his headship of the Oxford School of Geography.

In the late nineteenth century, with the help of the Royal Geographical Society, Oxford had appointed a Reader in Geography (H.J. Mackinder, in 1887), and has set up a School of Geography (in 1899). The early years of the school were distinguished and both Mackinder and Herbertson were major figures in British and international geography. However, by the 1920s the status of geography was low, and there was some risk in the 1920s that it might be abolished. One member of the board of studies that supervised geography was F.A. Lindemann (later Lord Cherwell), and he termed the discipline 'a kindergarten subject'. The colleges, with the exception of Hertford and Jesus, had little use for the subject, and the school itself had a very small staff. There were only four – J.N.L. Baker, Miss McMunn, K.S. Sandford (a geologist) and W.G. Kendrew (a climatologist) – and they were not all employed full-time. Mason arrived, therefore, at the time when things were at a very low ebb. He needed to attract undergraduates, to develop the college base, to obtain a larger and better-paid staff, to provide appropriate facilities for lectures, and to build up a research programme. By the time of his retirement he had achieved most of these objectives and his headship of the school was a period of immense progress in spite of the lack of cooperation he sometimes received from J.N.L. Baker. Between 1932 and 1939 the number of undergraduates increased from a mere handful to between 170 and 180 reading for Honours or for higher degrees. He built and equipped a large new lecture theatre. He obtained research scholarships from the Drapers Company to support three young research workers, including Mary Doveton (who worked on Swaziland), A.F. Martin (who worked in Newfoundland, and subsequently became a lecturer in the school and Fellow of St John's College), and Robert Steel, who went on to a distinguished academic career in Oxford, Liverpool and Swansea. After the war, as his obituary notice in *The Geographical Journal* (1976) recalls, 'Mason threw himself at once into the task of reconstructing his department, adapting it to post-war conditions and attitudes, and expanding its influence and activities. He was an efficient administrator and his occasionally rather gruff, military manner concealed considerable skill and persuasive powers in dealing with the complex bureaucracy of an ancient university'.

To conclude, Mason like many of the founders of the discipline in Britain, was not by training an academic geographer, and nor, indeed, was he an academic. He was essentially a practical surveyor and military man who, through the influence of the Royal Geographical Society, was introduced into academic life relatively late in his career. By the time he reached Oxford he had already had a distinguished career as a surveyor, soldier and explorer, and had made some underrated contributions to the study of glaciers and natural hazards in the mountains of High Asia. While at Oxford, Mason managed during the Second World War to mobilize a team to produce intelligence materials, including the Admiralty geographical handbooks. On either side of the war he transformed the nature and size of the Oxford School of Geography.

Bibliography and Sources

1. REFERENCES ON K.J. MASON

Goudie, A.S., Jones, D.K.C. and Brunsden, D., 'Recent fluctuations in some glaciers of the west Karakoram Mountains, Hunza, Pakistan', in Miller, K.J. (ed.) *The International Karakoram Project*, Vol. 2, 1984, 411–55.

Hewitt, K., 'European science in High Asia: geomorphology in the Karakoram Himalaya to 1939', in Tinkler, K.J. (ed.) *History of Geomorphology*, Unwin Hyman, London, 1989, 165–203.

Obituary notices in *The Times*, 3 June 1976, p. 19, and *The Geographical Journal*, Vol. 142, no. 3 (1976) 567.

Unpublished autobiography of K.J. Mason (ms in the School of Geography, University of Oxford).

2. BIBLIOGRAPHY OF WORKS BY K.J. MASON

1914 'The Indo-Russian Triangulation connection, 1911–1913', *Geographical Journal*, Vol. 43, 664–72.

1914 'Examination of certain glacier snouts of Hunza and Nagar', *Records, Trigonometric Survey of India*, Vol. 6, 49–51.

1926 'Movements of Indian glaciers', *Geographical Journal*, Vol. 68, 57–82.

1927 'The Shaksgam Valley and Aghill Range', *Geographical Journal*, Vol. 69, 289–332.

1929 'Indus floods and Shyok glaciers', *Himalayan Journal*, Vol. 1, 10–29.

1929 'The representation of glaciated regions of maps of the Survey of India', *Professional Paper No. 25*, Survey of India, Two Parts (Dehra Dun).

1932 *The Geography of Current Affairs*, Inaugural Lecture delivered before the University of Oxford, 15 November 1932, Oxford University Press.

1935 'The study of threatening glaciers', *Geographical Journal*, Vol. 85, 24–35.

1939 (with I.H.L. Grant) 'The upper Shyok glaciers', *Himalayan Journal*, Vol. 12, 52–63.

1955 *Abode of Snow*, Rupert Hart-Davies, London.

1987 *Abode of Snow* (photo reprint edition), Diadem Books Ltd, London.

A.S. Goudie is Professor of Geography in the University of Oxford.

Chronology

1887	Born at Sutton, Surrey, 10 September
1905	Entered Royal Military Academy, Woolwich
1906	Commissioned into Royal Engineers
1909	Joins Survey of India
1913	Indo-Russian Triangulation
1914	Sent to Western Front
1915	Injured at Ypres
1917	Awarded Military Cross and married Dorothy Robinson
1926	Leads Shaksgam Expedition
1927	Awarded Founder's Medal of the Royal Geographical Society
1932	Appointed Professor of Geography in the University of Oxford
1953	Retires from Oxford Chair of Geography
1976	Dies in Berkshire, 2 June

Frank Gilbert Roe

1878–1973

Gary S. Dunbar

Photograph (undated) by Kenneth McAllister, reproduced through the courtesy of the University of Victoria Archives

Although he never claimed the title of geographer, the Canadian scholar Frank Gilbert Roe should certainly be admitted to the fraternity. An autodidact who had a long career as a farmer and railroadman, Roe was truly a *géographe malgré lui*, a keen observer of both the natural and the human scene. His great thirst for learning enabled him to overcome the material circumstances of his early life and to win acceptance into the fellowship of scholars. Although he was 51 years old when his first publication appeared, and 73 when his magnum opus, *The North American Buffalo*, was published, Roe never seemed to regret having spent so many years in sometimes arduous physical activity. Rather, he felt that he was fortunate in having been so close to the land and to the workers who did so much to fashion the landscape.

1. Education, Life and Work

Frank Gilbert Roe was born on 2 August 1878 in Sheffield, England. Although (or because) the family had very modest means, reading was a great passion, and young Frank had an early exposure to the classics of English history and literature. Not only did he remember the facts in these books but he imbibed deeply of their style as well. His formal schooling ended at the age of 11, when he announced his intention 'to go to work and be a man' after his father had lost his job. At first Frank earned 3 shillings and six pence for a 44-hour week, but he managed to put aside a shilling a week and so was able, five years later, to pay his own passage from Liverpool to Winnipeg. Poverty left an indelible mark on Roe, but he never used it as an excuse; rather, he always seemed to be able to find positive benefits in his early experiences. He never wallowed in self-pity but, instead, prided himself on his ability to make his own way. He had real affection for his homeland even though conditions there had forced his family to emigrate. As he explained to the medievalist Julian Bishko in 1964:

> I have been living in Canada for 70 years, June 17 next, and I am here because of my parents' poverty. My father was married rather late in life, and when he began to lose pace in the labour market we children weren't old enough to support him. A very small legacy from a relative enabled us to go to Canada, after my father was turned loose by a man for whom he had worked for 24 years. There were no pensions then, and no 'Welfare State'. My father was foreman in a large grocery store, with a good salary of eight dollars a week. He supported his wife's parents for 13 years; which partly explains why such a man hadn't 'made provision for his old age', as his employer told him he should have done. ... We came to Canada because my parents didn't want to drop down below the poverty line in sight of those who had known them in better days. We could hide it away in Canada.

After a rather unpleasant Atlantic crossing in third-class accommodations on the SS *Parisian*, the Roe family disembarked in Quebec on 17 June 1894 and proceeded by train, first to Montreal and then to western Canada, reaching their destination, Red Deer, Alberta, about halfway between Calgary and Edmonton, on 5 July. The region had been described to them as 'the Garden of Alberta', and, since they had already known that Kent was the 'Garden of England' and the Lothians the 'Garden of Scotland', they felt that their new home would be equally fertile. Although the reality appalled them, they were down to their last $10 and so could not afford to exercise any real choice. The Roes took up a homestead 10 miles east of Blackfalds, Alberta and proceeded to erect a one-room sod house, which was replaced by a log cabin the following year. Before the log house was completed, however, the father died in 1895 from exposure to a winter storm. The rest of the family carried on, and in 1897 Frank took up a homestead of his own at the age of 19. There followed more than a decade of arduous physical labour. Before he could afford to buy barbed wire in 1905, he had to split thousands of fence rails ('2000 rails a rod long per winter was practically our regular tally') and erect and maintain the fences:

> Homesteading with not much more than one's ten fingers to start with was tough sledding ... I was just about getting to that stage where one stands to win the horse or lose the saddle, when two bad years together just broke my heart. I had a splendid crop hailed out in 1906 ... and snowed under in 1907. Life hardly seemed worth living.

With debts amounting to $700, a small fortune at that time, and with some unsympathetic creditors pressing him, Roe 'was most thoroughly fed up, and decided to drop it all and go railroading'. In August 1909 he moved to Edmonton and began his 34-year railroad career as a lowly engine wiper on the Grand Trunk Pacific Railway, which later became the main line of the Canadian National Railway through western Canada. He knew nothing of railroading and had already passed the age of 30, but he took to his new job with enthusiasm because, unlike farming, it offered steady work with certain rewards. His obvious intelligence and dedication soon earned him promotion to fireman and then ultimately (1919) to the rank of engineer. Roe's bearing and extensive vocabulary caused him to be elected by his fellow trainmen to take their grievances to the railway officials from time to time: 'I fortunately had a command of standard English, and didn't need to call the officials "sons of bitches" simply because I'd no other words I could think of.' He was never ill at ease in any stratum of society.

Edmonton offered far greater cultural resources to Roe, even though he began his career in the city by working 13-hour shifts seven nights a week. Always an avid reader, he now had some extra money with which to buy books, and he made full use of the municipal, provincial legislative, and university libraries. He made the acquaintance of a wealthy old man who allowed him to borrow books from his large private library of historical works on North America. Roe read widely in the classical works of English literature and history and did not at first plan any writing projects of his own. He had a natural critical sense, quickly discerning contradictions and inaccuracies in the various sources. His interests in the buffalo developed only slowly. He was struck by a statement that roads had their origins in animal trails, with American buffalo trails cited as proof. Roe was immediately sceptical of this assertion, and he set out to read systematically on the topic of animal trails in Europe and North America. He thought that the animal element in place names would provide significant clues, and he joined the English Place-Name Society in order to acquire its authoritative county studies. He put his own project on hold pending the publication of several of the society volumes that he thought would be essential to his work. 'So I suspended operations on the buffalo *trail*, and went after the buffalo *per se*.' His first publication was on buffalo trails ('The "wild animal path" origin of ancient roads', 1929), and his subsequent articles mostly concerned various aspects of buffalo history, culminating in the publication of his monumental book, *The North American Buffalo*, by the University of Toronto Press in 1951. Actually, the book was essentially completed some fourteen years earlier, when he began discussing its publication possibilities with various historians and editors. Some readers might have been put off by Roe's lack of academic credentials, others by the sheer bulk of the manuscript, but his persistence finally paid off when he got a favourable audience with the director of the University of Toronto Press, which had published Roe's earlier papers in the *Canadian Historical Review*. The book firmly established Roe's reputation – at the age of 73 – and he was given an honorary doctorate by the University of Alberta in the same year. A decade earlier, his name had been proposed for an honorary Master of Arts degree, but when professors objected to giving an honorary degree also to the premier of Alberta, it was decided that no honorary degrees would be awarded that year. He had won a host of friends at the university, notably the librarian, D.E. Cameron, and the professors and students in the departments of History, English Literature and Zoology. Had there been a department of geography, he would have been welcomed there as well. Many others had gained a great appreciation of Roe's intellect, including the Canadian naturalist Ernest Thompson Seton and the Texas folklorist J. Frank Dobie. (Roe's firm friendship with Dobie began at a conference in Lincoln, Nebraska in 1942.)

When the buffalo book got a warm reception from the scholarly community, Roe then produced a smaller manuscript on the Indian horse in North America. The book was published in 1955 by the University of Oklahoma Press. He prepared other book-length manuscripts, but no further books were published during his lifetime, apart from a second edition of *The North American Buffalo* in 1970. Roe did manage, however, to publish numerous articles in his remaining years, mostly in the *Queen's Quarterly* and the *Alberta Historical Review*. He was deeply touched by his election to Fellowship in the Royal Society of Canada in 1960.

Roe's publication output increased after his retirement from the railroad in 1943 and his removal to Victoria, British Columbia in the following year, but his physical activity did not diminish. He enjoyed remarkably good health and physical strength right up to the end of his long life. He explained his good fortune by saying that 'the man whose muscles have been hardened by real hard work has an easier job keeping them hard and himself in the resultant physical condition than the man with "exercise" muscles only ... I have never had to counteract the evils of a sedentary occupation'. For this reason, he never felt the need to engage in athletic activities, except for his love of distance swimming. In 1962 he told me that he swam across the Thames (and back) just above London less than four years earlier. At home on Vancouver Island he always had a lot of work to do to maintain his house and a large area of lawn and garden. For firewood he cut windfall trees in the adjacent woods with a one-man crosscut saw, and he claimed (1957) that 'I believe I can drive a sharp axe into a tree as deep as ever I could'. He always advocated the necessity of indoor hobbies as well as outdoor work, certainly in the fierce winters of Alberta but also in the rather more benign climate of British Columbia, taking as his motto that of Shakespeare's contemporary Horace Vere: 'All summer in the field; all winter in the study.' Along with his indoor activities of reading, writing letters, and preparing articles and books for publication, Roe also wrote reams of poetry, but never with the idea of offering any of it for public view.

Roe died at his home in the Cadboro Bay section of Victoria on 11 April 1973 at the age of 94. He was married twice, his second wife preceding him in death by four months, but he had no children.

2. Scientific Ideas and Geographical Thought

Whatever else Roe might have been called – or might have called himself – he was a geographer, specifically an historical geographer. His interests in re-creating past landscapes and in charting geographical change over time put him in the company of the geographers of his era, and it might even be said that he was ahead of his time in his ability to view these matters through different cultural lenses. He was aware, for example, of the different environmental perceptions of Europeans and Native Americans, of Canadians and Americans, of eastern and western Canadians, of farmers and ranchers, and of rural and urban people. Roe had all the attributes of a successful professor of geography or history – an exceptional memory, a keen analytical mind, a ready wit, persuasive logic, and strongly held opinions, all combined with a democratic and open manner that gave him immediate rapport with people in all walks of life. If he had been born a generation later and had had the advantages of a university education, he undoubtedly would have become a professor. As it was, he took a half-century detour before becoming accepted as an equal by university scholars.

Roe had a marked advantage over most academics: he knew at first hand what he was talking about. He had felled trees; he had split rails; he had walked behind a plough; he had shovelled coal for steam locomotives; he had even been called up before a local magistrate for letting a prairie backfire get out of control. When he wrote an article about the Alberta wet cycle of 1899–1903 for the journal *Agricultural History* (1954), he was writing from intimate personal knowledge; he had lived through that cycle – as well as previous and subsequent dry or cold summers and times of devastation by wind, fire, and insect infestations. Although he did not get beyond the borders of Alberta for the first 19 years after his arrival in 1894, he subsequently managed to do a fair amount of travelling, with numerous train trips across Canada, several journeys back to England, and also excursions to the USA ('about 30 States in three different journeys'), Mexico, and the West Indies.

Roe's best-known work is the book *The North American Buffalo*, whose prolonged gestation would have discouraged most other authors. Although he came to the Canadian prairies too late to observe the large bison herds that roamed throughout the western plains, he was able to interview older residents whose memories were still fresh, and his photographs captured some of the essence of earlier times. One of the most remarkable sights on Roe's initial train trip from Winnipeg to Alberta in 1894 was 'a long stack of bleaching buffalo bones alongside the track' at Maple Creek, Saskatchewan, but his interest in writing something about buffalo was not ignited until some three decades later. His initial scholarly interest had been in old roads and trails in England, specifically in their alleged origin as animal trails, but he subsequently turned to a concentration on buffalo trails in North America and then to the buffalo itself. He always read with a critical eye, and his scepticism was aroused by the facile assumption that the first human trails followed animal trails. His greatest criticism was levelled at the work that had been universally accepted as the most authoritative treatment of buffalo, 'The Extermination of the American Bison, with a Sketch of Its Discovery and Life History' (1889), a 180-page Smithsonian publication written by William T. Hornaday, whom Roe later met. Although Hornaday had had the advantage of writing at a time just after the near-annihilation of the great buffalo herds in the third quarter of the nineteenth century, Roe took serious exception to many of Hornaday's generalizations and produced a more exhaustive (nearly a thousand pages) and withal a more historically accurate book on the subject. New palaeontological, archaeological and historical data have been uncovered since 1951, some of which Roe took into account in the second edition in 1970, but the book has remained the foundation stone of buffalo scholarship, the obvious starting point for further research on the subject. The range of the buffalo in the historic period has been shown to extend farther southwestward than Roe was at first aware, and the geographer Erhard Rostlund has demonstrated the deep penetration of buffalo into the southeastern USA. A.J. Ray, D. Wayne Moodie, J. Gordon Nelson and Jerry McDonald are other geographers who have added to our knowledge of the buffalo, but no one has yet produced a work to match the scale of Roe's.

For acute commentary on Roe's scholarship and methodology, as demonstrated by the buffalo book, I turned to the doyen of Canadian historical geographers, John Warkentin, who said, in 1996:

> Roe's methodology was of great interest to me and a practical lesson, because his scholarly procedures were laid out in full sight right before one, almost in skeletal form. He used all possible sources meticulously and critically questioned the accuracy and validity of everything, and was careful not to go beyond the evidence at hand in making generalizations. There was an immediacy and directness to his analysis, because immediately as he presented the observations of the countless observers he quoted, he also bluntly evaluated the reliability of what they reported, but always fairly, giving reasons and placing the observations in a comparative context. One is comfortable with the bluntness and constant evaluations because there is never any doubt that though one is reading a strong willed man of independent mind and a sharp turn of phrase, the entire huge book (or essay as he refers to it!) is above all an intellectual inquiry. Despite all the detail, the quotations, and the constant judgments, there is a pervading sense of true curiosity, of wanting to know, of being on an intellectual voyage of discovery with the author, and this gives an inner vigour and dynamic to the book that will always make it fascinating to read.

Also in 1996 Charles Julian Bishko, who carried on an extensive and intellectually rich correspondence with Roe over a period of two decades and who is known to geographers for his publications on medieval Iberia and its later extensions in Latin America, gave me his retrospective views on Roe's buffalo and horse books:

> In his massive *The North American Buffalo*, he subjects to astringent critical analysis the abundant but often contradictory eyewitness testimonies while surveying with balanced judgments the variant types of Plains and Wood buffalo, their habitat and distribution, the migration routes of the immense northern and southern herds, the controversial but consistently huge demographics, and the final American and Canadian exterminations. In *The Indian and the Horse*, Roe similarly scrutinizes long-current misconceptions regarding the impact upon the Plains Indian tribes of the acquisition of the European horse between the later 17th and 19th centuries. From meticulous review of the evidence adduced for fundamental changes in hunting patterns, migratory movements, transport and mobility, and social, economic and psychological patterns, he concludes that the horse, while indubitably extending and enriching Indian life, did not in fact revolutionize the age-old Plains tribal structures, affecting mentality and outlook more profoundly than material culture. Both of these substantial studies, exhibiting expert control over recalcitrant sources, lively dissent from accepted conventions, and high originality, remain indispensable to understanding the animal and aboriginal societies of the Great Plains.

The horse book was to attract more criticism than the earlier work on the buffalo, probably because anthropologists know more about the Indian horse, and certainly about Indians generally, than they do about buffalo. And maybe the far greater length of the buffalo book might have exhausted some readers and stifled their intended criticism. Anthropologists criticized Roe's statements about the place of buffalo and horses in Indian cultures, geographers wished for maps and a more extensive treatment of the physical geography of the Great Plains, and biologists noted the less than satisfactory treatment of genetics, but, all in all, most reviewers praised Roe's works for their general usefulness as compendia of knowledge on their subjects. Inevitably, there was some grumbling about Roe's prolixity and bibliographical overkill, traits that would not appear in books published today, simply because publishers would not allow them. Who could manage to sneak a musical notation of a coyote's howl into a footnote, as Roe did in his horse book?

Apart from the books, geographers would find much of interest in Roe's articles on other subjects, such as English fields, forests, and place names, railroad building in western Canada, and prairie farming. There are great similarities among all his published works: heavy documentation; frequent allusions to biblical, classical and Renaissance literature; a rather quaint writing style reflecting his early devotion to English historical and literary works of the sixteenth to nineteenth centuries; strong opinions supported by personal experience and wide reading; and no attempt to conceal the author's curiosity and enthusiasm. Roe was a rare individual indeed!

3. Influence and Spread of Ideas

Although he lacked students and was never in a position to bestow preferment on anyone, Roe did have the influence that comes with publication of articles and books. His ideas will continue to be spread among people who relish sound historical scholarship. All subsequent writers who have wanted to add to the story of the buffalo and horse in North America have had to begin with Roe's books. Of increasing value to researchers will be Roe's personal observations and photographs, which will be even more useful than his citations of earlier writers. His style and method should also be of interest to neophyte scholars, even though they will not want to copy them directly. If any detractors criticize his empiricism, what will they offer to replace it?

Perhaps even more influential could be Roe's life history. It would be impossible to pattern one's life after his, but it is a humbling and salutary experience to try to see the world through his eyes. Today, it seems, book scholarship and direct field experience are not so highly prized by younger geographers, and few would want to adopt Roe's dictum about spending summers in the field and winters in the library, but it may not be too late to instil respect for the ways of their intellectual ancestors.

Acknowledgements

In writing this essay, I profited immensely from communications with my friends Professor (Emeritus) Charles Julian Bishko of the University of Virginia and Professor (Emeritus) John H. Warkentin of York University. Their knowledge of the life and work of Frank Gilbert Roe is superior to mine, but the (not unpleasant) task of writing this biographical sketch has fallen instead to me. I should also like to thank Mr Dietrich Bertz, Archives Assistant in the University of Victoria Library, and Professor (Emeritus) W.A. Fuller of the University of Alberta.

Bibliography and Sources

1. ARCHIVAL SOURCES

The Frank Gilbert Roe Collection in the University Archives in the University of Victoria Library, Victoria, British Columbia, contains correspondence (incoming and outgoing), publications, manuscripts and typescripts, photographs, maps, personal documents, and a phonotape interview with Roe. There is a 119-page finding aid to the collection. Perhaps the most valuable exchange of letters was between Roe and C.J. Bishko from 1953 to 1973 (Roe Collection in the University of Victoria and Bishko Papers in Alderman Library of the University of Virginia). In 1996 Professor Bishko supplied the Roe Collection with photocopies of letters that they did not previously have. Professor Bishko's letters to me (1995–7) concerning Roe have been of crucial importance in writing this essay. I have acquired copies of J.H. Warkentin's letters (1960) to Roe from the University of Victoria, and Professor Warkentin has recently (June 1996) sent me some further reflections on the significance of Roe's buffalo book. I myself exchanged two letters with Roe in the spring of 1962.

2. WORKS ABOUT FRANK GILBERT ROE

1974 Stanley G.F.G., 'Frank Gilbert Roe, 1878–1973', *Proceedings of the Royal Society of Canada*, Series 4, Vol. 12, 63–5.

n.d. 'Frank Gilbert Roe, 1878–1973', pp. 1–3 of the finding aid to the Roe Collection, University of Victoria Archives, by an anonymous author.

3. FRANK GILBERT ROE: A SELECTED BIBLIOGRAPHY

1929 'The "wild animal path" origin of ancient roads', *Antiquity*, Vol. 3, 299–311.

1934 'The extermination of the buffalo in western Canada', *Canadian Historical Review*, Vol. 15, 1–23, 213–18.

1935 'The Red River hunt', *Transactions of the Royal Society of Canada*, Vol. 29, 171–218.

1936 'Buffalo and snow', *Canadian Historical Review*, Vol. 17, 125–46.

1936 'The crooked field', *Antiquity*, vol. 10, 325–40.

1936 'An unsolved problem of Canadian history', Canadian Historical Association, *Annual Report*, 65–77.

1937 'The numbers of the buffalo', *Transactions of the Royal Society of Canada*, Vol. 31, 171–203.

1939 'The winding road', *Antiquity*, Vol. 13, 191–206.

1939 'From dogs to horses among the western Indian tribes', *Transactions of the Royal Society of Canada*, Vol. 33, 209–75.

1939 'Buffalo as a possible influence in the development of prairie lands', *Canadian Historical Review*, Vol. 20, 275–87.

1944 'White buffalo', *Transactions of the Royal Society of Canada*, Vol. 38, 155–73.

1946 'Early opinions on the "fertile belt" of western Canada', *Canadian Historical Review*, Vol. 27, 131–49.

1951 'Address to the Graduating Class, October 20, 1951', *The New Trail* (University of Alberta), Fall 1951, 166–72. Convocation address delivered by Roe when he was awarded the honorary degree of Doctor of Letters from the University of Alberta. It is a rambling discourse that displays his humanity and wide learning. It is personal without giving much autobiographical data. To a remarkable degree, almost everything that Roe ever wrote contained personal elements, thanks to his keen powers of observation and excellent memory.

1951 *The North American Buffalo: A Critical Study of the Species in Its Wild State.* University of Toronto Press, Toronto. Second edition, 1970. Among the reviews, I should like to draw particular attention to two: by Luna Leopold in the *Geographical Review* (January 1953) and Richard Glover in *Saskatchewan History* (Spring 1952).

1952 'Early agriculture in western Canada in relation to climatic stability', *Agricultural History*, Vol. 26, 104–23.

1954 'The Alberta wet cycle of 1899–1903: a climatic interlude', *Agricultural History*, Vol. 28, 112–20.

1955 *The Indian and the Horse*, University of Oklahoma Press, Norman.

1955 '"Forests" and woods in mediaeval England', *Transactions of the Royal Society of Canada*, Vol. 49, 67–93.

1956 'What is a "field"?', *Transaction of the Royal Society of Canada*, Vol. 50, 21–37.

1957 'Western penetration of the historic buffalo in the Upper Bow River Valley', *Alberta Historical Review*, Winter, 21–4.

1958 'The old log house in western Canada', *Alberta Historical Review*, January, 1–9.

1958–9 'Pilgrim at Glastonbury: history's stones', *Queen's Quarterly*, Vol. 65, 650–63.

1959 'A week in southern Ireland', *Queen's Quarterly*, Vol. 66, 450–67.

1960 'Buffalo trails and fur posts', *Queen's Quarterly*, Vol. 67, 449–61.

1961 'Some historical evidence on the earlier physiography of the North American prairie', *Transactions of the Royal Society of Canada*, Vol. 55, 9–35.

1964 'Edmonton a century ago', *Alberta Historical Review*, Vol. 12, 10–16.

1970 'The sod house', *Alberta Historical Review*, Vol. 18, 1–7.

1972 'A day in court', *Alberta Historical Review*, Vol. 20, 1–5.

1982 *Getting the Know-How: Homesteading and Railroading in Early Alberta*, edited by James Patrick Regan, foreword by Lewis G. Thomas. NeWest Press, Edmonton. This is Roe's autobiography, covering the years 1894–1944. The editor has made 'an amalgam of two historical/autobiographical essays', one written in the mid-1940s and the other 'possibly dating from the early 1960s'. Four chapters were previously published in the *Alberta Historical Review*.

Gary S. Dunbar is Professor (Emeritus) of Geography in the University of California, Los Angeles. His current address is 13 Church Street, Cooperstown, New York 13326, USA.

Chronology

1878	Born 2 August 1878 in Sheffield, England
1894	Family emigrated to Canada, arriving in Red Deer, Alberta, on 5 July
1894–1909	Farmed, first on family farmstead, and, beginning in 1897, on his own
1909	Began his railway career as an engine wiper on the Grand Trunk Railway in Edmonton, Alberta
1924	Joined English Place-Name Society
1929	First published article (in *Antiquity*)
1943	Retired from the Canadian National Railway
1951	*The North American Buffalo* published by the University of Toronto Press
1951	Honorary Doctor of Letters, University of Alberta
1952	Learned to type
1955	*The Indian and the Horse* published by the University of Oklahoma Press
1960	Fellow of the Royal Society of Canada
1973	Died 11 April in Victoria, British Columbia

John Lort Stokes

1811–1885

Marion Hercock

National Maritime Museum London

John Lort Stokes RN was a surveyor and naval officer in the British Admiralty's Hydrographic Service. He was the epitome of the nineteenth-century British naval surveyor, in his length of service and dedication to duty, and in the professionalism shown by his scientific and geographical approach to surveying and exploration. He exemplifies the inquiring mind and broad interests and skills of a geographer. Stokes' contribution to the geographical knowledge of Australia, and his detailed charts of sections of the coast, stand out as milestones in the exploratory history of the country. His role in the exploration of the north-west and north of Australia is exceptional in the history of Australian maritime exploration, as he carried out exploratory surveys well inland. Stokes' achievements are in the great tradition of Royal Navy hydrography, a tradition based on the provision of accurate charts and navigational information, on a world-wide basis, for naval purposes and for the British mercantile marine. The experience of scientific exploration gained from serving with Phillip Parker King and Robert FitzRoy on surveys of the South American coast provided a foundation for Stokes' own commands in Australia, New Zealand and Britain. The association with contemporary explorers and scientists, including Charles Darwin, put Stokes at the cutting edge of geographical investigation in the mid-nineteenth century. His major published work reveals one who was interested in landscape, geology, vegetation and wildlife, native peoples and colonial society – in short, all that should concern the geographer. Stokes earns a place in the annals of Australian exploration and of the British Empire, and in the history of geographical research, through the quality of his work in the field.

1. Education, Life and Work

John Lort Stokes was born on 1 August 1811 at his father's family estate, Scotchwell, near Haverfordwest, in Pembrokeshire. The second son of Henry and

Anne (née Phillips) Stokes, he was named Lort after his mother's father. The family name, Stokes, had been assumed by his grandfather, Mr John Rees, who had had some associations with trade. It was at Scotchwell that Stokes' life-long interests were nurtured – field sports, and the countryside and its people.

His education in surveying began a year after he had volunteered for naval service at the age of 13. Following entry into the Royal Navy as a first-class volunteer in 1824 on board the *Prince Regent*, the young Stokes joined the brig *Beagle* as a midshipman in 1825. For 18 years Stokes was to serve on the *Beagle*, during which he would see the ship in a variety of circumstances and situations, from Tierra del Fuego to the north-west of Australia. His training in the techniques of hydrographic survey was undertaken during the expedition led by Captain Phillip Parker King to survey the coasts of South America. Navigation and hydrographic matters were not the sole concern of the expedition's naval officers. The Admiralty had also instructed King and his officers to make collections 'of such objects of Natural History as may be new, rare or interesting'. Thus, Stokes was introduced to the range of biophysical subjects considered by scientists on expeditions of research and discovery. It was on this 1826–30 survey that Captain Robert FitzRoy assumed command of the *Beagle*; this being upon the death of Captain Pringle Stokes (no relation to John Lort) in 1828. FitzRoy had some involvement in the appointment of Charles Darwin to the *Beagle*'s next expedition.

In 1831 HMS *Beagle* was recommissioned, under FitzRoy, a first-class hydrographer and pioneer meteorologist, for a second hydrographic and general scientific expedition to the Strait of Magellan and South America. Stokes was appointed mate and assistant surveyor on one of the most significant expeditions in the history of scientific thought. The achievements of the expedition and its naturalist, Charles Darwin, owe much to the effectiveness of the *Beagle*'s officers as a scientific research team. Stokes' devotion to survey and the quality of his work, typical of the tradition of the surveyors of the period, was remarked upon by his commander. In 1833, at Monte Video, Captain FitzRoy wrote in a private letter to the Admiralty Hydrographer: 'Two things I am *very* anxious about. The application I have made to you for Stoke's promotion ... Stokes is my constant, and my oldest, ally in the *Beagle*. His *work* appears to *me* to support my earnest request' (Archives, Hydrographic Department, Taunton, cited by Keynes, 1979). As the expedition had been testing Beaufort's ideas and approaches to the gathering and reporting of hydrographic data, FitzRoy added: 'tell me exactly what *you* think of the documents now sent to you. I have tried much to put in practice your ideas, but as we are all beginners, what appears right to me, may not suit good judges' (Archives, Hydrographic Department, Taunton, cited by Keynes, 1979). FitzRoy was among the generation of nineteenth-century naval officers who were not only career servicemen, but also, increasingly, professional scientists. Stokes, inculcated with the scientific approach to hydrographic work and exploration, would perpetuate the professional standards of rigour and precision. Stokes' capacity for, and competency in, survey work, and his 'steady and able head' on the expedition earned him a promotion to lieutenant, on the return of the *Beagle* to London in January 1837.

The following month, the *Beagle* was assigned to a third major expedition, under Captain John Wickham, to survey the Australian coastline at the north-west, Bass Strait and Torres Strait. Again, Stokes served as assistant surveyor. The Admiralty was particularly interested in the discovery and exploration of large rivers on the Australian coast, as it was thought that one such river could flow from an inland sea. The concept of an opening into an inland sea had been a feature of the maritime and overland exploration of Australia, since the commission of Captain Matthew Flinders in 1801 to survey the coastline. The *Beagle*'s third great expedition was to investigate the north-west coast for possible outlets from an inland sea; elsewhere she was to concentrate on hydrographic and meteorological work, with only general attention to geology and natural history. During the six-year expedition, Stokes became the first naval surveyor to carry out substantial inland journeys in Australia as adjuncts to the marine survey work. The inland explorations generally began with a journey upriver in a whale-boat, from which scouting parties went ashore for short distances, seeking high ground or following valleys. In this manner the surveyors were able to penetrate the inland, extending, in the case of the Victoria river in northern Australia, up to 140 miles (225 kilometres) from the sea. The expedition's two circumnavigations of the country, and the many inland expeditions, contributed to the geographical knowledge of Australia, through the collection of sufficient data to lead to the abandonment of the notion of an inland sea, and through the British discovery, and subsequent naming, of major rivers and inlets. Stokes is credited with the discovery of the Port of Darwin and the FitzRoy, Albert and Flinders rivers. His enthusiasm for exploration and his interest in his work was such that not even a near-fatal spear wound received in northern Australia in 1839 deterred him from continuing a career in the Hydrographic Service. For the rest of his life, Stokes would suffer from the effects of the wound to his lung. In contrast, the debilitating conditions of the tropical north of Australia contributed to the deterioration of Captain Wickham's health, to such a degree that he had to give up his command. Wickham was invalided out of the navy in 1841, handing over command of the *Beagle* to Stokes, who completed the rest of the expedition, returning to England in 1843. This return completed 18 years of service on one ship, an association and career path unparalleled in the Royal Navy.

After finalizing the charts of the Australian expedition for the Admiralty and commercial production by the London cartographer John Arrowsmith, Stokes was elected a Fellow of the Royal Geographical Society in 1845. In the following year, Stokes was advanced to post rank, and two volumes about his experiences on the *Beagle*'s last expedition were published, by order of the Lords of the Admiralty. The book, entitled *Discoveries in Australia*, was dedicated to Robert FitzRoy and included sketches made by Lieutenants Gore and Fitzmaurice, additional papers by Benjamin Bynoe, the ship's surgeon, and numerous footnotes. Among the other contributions were a list of the birds collected during the voyage and identified by John Gould, and detailed descriptions of six species of fish taken on the expedition. Several maps published by Arrowsmith were featured in the Appendix, as well as supplementary descriptions of Australian insects by the British Museum, and of reptiles by the land explorer, John Edward Grey. John Wickham contributed a section on the winds and weather of the western and northern coasts of Australia. Although written for the general public of Victorian Britain, the book remains a useful reference for geographers, owing to the detailed descriptions of the biophysical environment, and the activities of the explorers. For example, Stokes' descriptions of an excursion up the Fitzroy River in north-western Australia accurately depict the character of one of the largest river systems in Australia:

> Hastening back to the boats we pushed on, but were some time getting to the end of the reach, the shallowness of the water rendering our advance difficult and tedious; entering at length the next, which tended S.W. for about half a mile, the river gradually widened out until it attained a breadth of about half that space. An extensive flat of sand fronted the eastern bank, which was very low, and though now dry, bore undoubted marks of being not infrequently visited by floods. The western

> bank of the next reach was low and broken, evidently forming a group of low grassy islands when the river is in a higher state. Some yellow sandstone cliffs, from ten to sixteen feet in height, formed the opposite bank of this reach, which extended barely a quarter of a mile... (Stokes, 1846, Vol. I, p. 147)

In 1847 Stokes was formally promoted to captain, and given the command of the five-gun paddle-wheel steamer HMS *Acheron*, with instructions to survey New Zealand. This work was to update the only charts then in use, which dated from the time of Captain James Cook's work in the 1770s. *Acheron* was the first surveying vessel with auxiliary power to be sent to Australasian waters, and proved to be of particular use on the New Zealand coast, where high winds prevail. Although the death of his wife, on the journey out to New Zealand, greatly affected Stokes, he spent from 1848 to 1851 carrying out the survey of the coast of New Zealand's Middle Island. (Today this island is known as the South Island, being one of the two largest islands of New Zealand. The other is known as North Island.) In addition, he surveyed parts of the New South Wales coast, all to the satisfaction of the Admiralty Hydrographer. A paper, presented to the Royal Geographical Society in 1851 from the Admiralty, was based on that travel and survey work. The editor of the Society's journal gives some indication of the calibre and significance of Stokes' Middle [South] Island survey: 'The distinguished character of the officer under whose superintendence this great work is being effected will stamp it with the highest authority, and few operations could have been undertaken of more importance to an insular colony' (Editors' note, *Journal of the Royal Geographical Society*, Vol. XXI (1851), 29).

After his return to Britain, Stokes was shore-based on half-pay at Scotchwell for the 1850s, and served on the council of the Royal Geographical Society from 1856 to 1857. Many retired naval officers served on the council, and the development of geography as an integrative discipline owes much to this, and other links between the Admiralty and the society. The prominence of naval officers in the society followed from their service careers and training, and they were well placed to add to geographical knowledge, being able to observe the global environment at sea. These qualities were commented upon by the president of the society (a naval officer) in 1850:

> But of all classes of individuals, to whom we may at present look for the promotion of some or other of the various ramifications of geography, none have such frequent and available opportunities as naval and military officers. Visiting as they do, in the public service, all the most interesting spots of the world, the field is laid invitingly before them; and it is gratifying to know, that the requisite scientific attainments for such operations and researches are already widely diffused among them, and are daily becoming more so. (W.H. Smythe, presidential address to the Royal Geographical Society, 1850)

The period that Stokes spent ashore was briefly broken by surveying work on the south coast of Devon from 1860 to 1863. Semi-retirement with the rank of rear-admiral in 1864 led to the remainder of his life being spent ashore. For the next 21 years Stokes was involved in estate and country life, enjoying the field sports, and serving as a magistrate for Haverfordwest and Pembrokeshire. He continued to maintain contact with various associates, clubs and societies, and corresponded on issues of maritime and national interest. In 1876, he was finally promoted to the rank of full admiral, and retired officially from active service in the following year.

2. Scientific Ideas and Geographical Thought

The contribution of Stokes to the geographical knowledge of Australia, both official and popular, is probably most apparent to the modern Australian reviewer of his major published work and his charts. If he is viewed as a field researcher, as one of a team of scientific researchers, a link in the chain of ideas, and part of a scientific movement, his place in the history of geography becomes clear. That place is reinforced by Stokes' part in the exploration of Australia. In his work he applied scientific principles to geographical explanation and recording, and left exemplary geographical field descriptions. He was a practical field scientist rather than an innovator of concepts.

To view Stokes as a geographer it is necessary to go beyond the assessment of the links between the Royal Navy, the Royal Society and the Royal Geographical Society as part of the British imperial endeavour under the auspices of *Pax Britannica*. Imperial aggrandizement and commercial expansion aside, the advancement of geography as a discipline was stimulated and projected by the intellectually productive links between the Admiralty and the scientific societies. Throughout Stokes' career, advances were made in hydrographic and meteorological theory, techniques, methodology and reporting. The role of the naval surveyor at sea was critical in the development of a uniform approach to coastal navigation, an approach which was scientifically rigorous, while having regard for the concept of natural processes operating at different scales. Stokes' work finds a place in the annals of the relationship between science, geography and the Admiralty, especially the Hydrographic Service.

The geographer's concept of climatic and coastal processes operating at different scales, global, regional and local, was well established in Admiralty hydrographic thought since the great transequatorial explorations of the eighteenth century. The association between geography and the Admiralty was fostered by Sir John Barrow, Second Secretary to the Admiralty for 40 years from 1804. Barrow, with the Admiralty Hydrographer, Sir Francis Beaufort, and several other gentlemen had constituted the Geographical Society of London, under the royal patronage of William IV, in 1830. Among the original 460 members were many distinguished surveying officers. In the following year, the Admiralty established a scientific branch, made up of the Hydrographic Department, the Royal and Cape Observatories, the Nautical Almanac Office and the Chronometer Office. Scientific expeditions and 'rewards' for special services or investigations were included in the budgeted items for expenditure. The part played by Beaufort (Hydrographer 1829–55) in fostering a close association between the Admiralty's Hydrographic Survey Department and the surveyor afloat was central. Both the department and Beaufort were active in the scientific fields of astronomy, meteorology and tidal theory. Looked upon as a fatherly mentor, as well as a scientist and captain, Beaufort maintained regular correspondence with his surveying commanders. He watched, and saw to, Stokes' development as a surveyor, from FitzRoy's 1833 recommendations for Stokes' promotion, to the confirmation of his command of the *Beagle* in 1842:

> I congratulate you on your appointment to the command of the Beagle and on the fair scope that is now open to your talents and enterprise – and I very sincerely hope that the discovery of some important river, the successful process of your survey, and the zealous effort of your officers and crew will crown your appointment and distinguish the onset of your career. (Beaufort to Stokes, 30 September 1841; cited by Hordern, 1989, p. 284)

The practical experience of marine surveyors led to the refinement of hydrographic survey methodologies and theory, as well as contributing to the greater understanding of climatic and coastal processes. Stokes carried out his orders according to the standards and methodologies laid out in the official surveying manual, *A Treatise on Nautical Surveying*, compiled by Commander Edward Belcher in 1835. This manual provided the surveyor with details of how to organize a survey, with descriptions of the methods and instruments in use. Surveyors were expected to report on the nature of the shore, topography, views, soundings, deep soundings, currents, magnetism, wind and weather, geology, sailing directions, and natural history. A surveyor would have to complete a journal, and detail the charts made during the survey, including the scale used and nomenclature. The approach was essentially scientific because it was replicable, and geographical because it studied the elements of the biophysical environment as a whole, in relation to each other.

The capacity for the naval surveyor to take a geographical view, encompassing all elements of the biophysical environment and its human aspects, is illustrated by Stokes' *Discoveries in Australia*. The inclusion of particular subject material in the book was based on the example and format of the *Nautical Magazine* and the *Journal of the Royal Geographical Society*. The *Nautical Magazine* and, after 1849, the *Manual of Scientific Enquiry*, which were greatly supported by Beaufort, featured reports by the naval surveyors. In addition, the *Journal of the Royal Geographical Society* was another medium for accounts of Admiralty surveys. The prominence of the geographer's perspective in the practice of the Hydrographic Service is evident from the subjects covered by the Admiralty *Manual of Scientific Enquiry*. This manual featured discussions on astronomy, magnetism, geography, geology, mineralogy, meteorology, zoology and ethnology and, of course, hydrography. Thus, the links between the discipline of geography and the marine surveying carried out by the Admiralty existed at several levels, philosophical, theoretical and practical.

Although the *Beagle*'s 1830–6 expedition included Charles Darwin as the scientific observer, the naval officers still devoted time to observing and noting the geology and wildlife. FitzRoy commented that Stokes had made notes about the drinking habits of the tortoises of the Galapagos Islands. In addition, Stokes showed an interest in the appearance and customs of the natives of the places visited. The practice of observation and notation was well developed by the time Stokes visited Australia, and the accuracy of his descriptions contributes to the high quality of *Discoveries in Australia* as a geographical record. For example, the southern shore of Port Phillip Bay in Victoria is described as:

> a singular long narrow tongue of land, running out from the foot of the range of which Arthur's Seat is the most conspicuous point. I infer from the limestone prevailing in it, and containing shells of recent species, that it was once beneath its present level; in fact, that it stops up what was formerly a broad mouth of the bay, leaving only the present narrow entrance at the western extremity. Over its surface are scattered hills from one to two hundred feet in height, in the valleys between which was found some light sandy soil supporting at this time rich grass, and at various places a thin growth of Banksia, Eucalypti, and Casuarina, all stunted and showing symptoms of having been roughly used by the south wind. (Stokes 1846, Vol. 1, p. 261)

Vegetation, topography, soils and landforms are discussed briefly, yet precisely, as the relationships between them are also noticed.

The survey of Bass Strait, the complex stretch of water between Tasmania and the mainland of Australia, is considered to be Stokes' greatest achievement in surveying. Based on work carried out on sea and land, the survey involved mastering the difficulties of communication between a number of teams working in the Strait, as well as the treacherous regimes of wind and wave. Points for triangulation had to be established on land for readings to be made from sea and on land, and such points were also noted as to their suitability for the location of lighthouses. The survey took two months and covered the waters from the northern coast of Tasmania to the coast of Victoria (then part of New South Wales). Gaps in the existing charts were filled in with outlines, depths, heights, names, wind and tidal information, and courses. The southernmost tip of King Island is named Stokes Point, and the *Beagle* provides names for a reef, a spit, and a rock in the Strait. The near loss of the ship in fine weather in Murray Pass, a narrow stretch of water between Deal and Erith Islands of the Kent Group, to a racing tidal eddy, led to a warning for mariners on the chart.

The completion of the *Beagle*'s Australian survey added to evidence that the country's rivers did not flow from an inland sea. Stokes stated in *Discoveries in Australia* that the theory of an inland sea had long vanished from his mind, and his own impression

> is that instead of an inland sea, there is in the centre of Australia a vast desert, the head of which, near Lake Torrens, is not more than three hundred feet above the level of the sea. The coast being surrounded by hilly ranges, the great falls of rain that must occasionally occur in the interior, may convert a vast extent of the central and lowest portion of the continent, into a great morass, or lake, which from the northerly dip must discharge its waters into the Gulf of Carpentaria, without possessing sufficient stability to mark either its bed or boundaries. (Stokes, 1846, Vol. 2, p. 12).

This view was not unreasonable, considering the knowledge about the interior at that time.

Stokes' flair for exploration inland, and the quality of his geographical interpretation, is shown in the report on the investigation into the location of Port Grey, near Geraldton in Western Australia. This report was prepared in December 1841, for the Surveyor-General of Swan River Colony, John Septimus Roe, himself a former hydrographic officer, who had also served under Phillip Parker King. The location, landforms, soils and vegetation are accurately described with precise measurements of bearings, altitude, temperatures and the course followed. Ethnographic comments on an Aboriginal grave and shelters, and an encounter with the natives, are included. The descriptions of the *Kwongan* sandplain vegetation, 'not a tree large enough to furnish building materials, being only small kinds of banksia, wattle and drooping gums – these grew only in the valleys' and the characteristic eucalypts 'of a dwarf and crooked description' are succinct, yet correct. Furthermore, from the details given, the course followed by the exploring party (the path taken by Stokes and his men) is replicable. For example:

> From an elevation on its S.W. side [the Chapman River] Mount Fairfax bore North 50° East, Wizard Peak South 58° East. From this point we proceeded one mile West over a dry, arid plain, covered with yellow and white everlasting flowers of small growth; a small patch of woodland, consisting of a species of wattle and a very small kind of gum, here delayed our hasty progress. (Report, 19 December 1841, *Exploration Diaries*, Battye Library, Perth)

The report must have satisfied the Surveyor-General, for he later offered Stokes a position as his assistant.

On the survey of the Middle [South] Island of New Zealand, particular exploration procedures were repeated. For example, the practice of taking a whale-boat upriver in order to investigate the interior was continued. The following extract illustrates that same quality of understanding, and description of the environment in a geographical manner, that Stokes exhibited in his descriptions of Australia:

> The New River was ascended in a whale-boat for nearly 30 miles in a N.½E. general direction. In that distance the land rose gradually 200 feet, by 3 steps, each change of elevation being attended by corresponding and somewhat dangerous rapids. The depth of the water varies from 2 to 8 feet, the width from 50 to 500 yards. The soil on either bank consists of a rich mould, and appeared clothed with trees or verdant pasture, as the stream wound through clumps of wood or swept across the open plain. (Stokes, 1856, p. 25)

During his time on the council of the Royal Geographical Society, Stokes presented a paper on the advantages of a steamship route to Australia, from Singapore via the Torres Strait. His arguments for this route, and against the route to the west and south of Australia, are based on his own experiences of the winds and waters, the interests of the Australian colonies, and his comprehension of the capabilities of steam transport in contrast to those of sail. In addition, he points out the commercial potential in expanding British transport links east of Singapore, the agricultural and developmental opportunities in Queensland, and the superior facilities of Sydney's harbour to those of Melbourne. Stokes closes the paper with an emphasis on the qualities of 'regularity, economy and speed, particularly homeward, of the Torres Strait route' compared with the western and southern route. His comments on the tranquillity of the route reveal his interest in landscape as well as his understanding of the realities of sea-travel

> As to the comparative comfort of the two lines there can be no question; on the one there will be a continual roll, with nothing but sea! sea! sea! – while the other offers placid waters and picturesque scenery through its greater part. (Stokes, 1856, p. 188)

In consideration of both physical and social aspects of the route, Stokes shows the comprehensive approach of a geographer. In addition, the paper is an example of the intellectual output generated by the relationship between the Admiralty and the Royal Geographical Society.

Although Stokes was clearly scientific in his work, his naming of landmarks was based on the names of patrons, friends and shipmates, and the events of the expedition. Port Darwin was named after one old friend, and the names of Disaster Inlet and Escape Cliffs echo the accidents and dangers of exploration. This practice did not conform to Beaufort's standards for geographical names, as the Hydrographer considered that a name should convey some idea of the sense of place, or make an allusion to the inhabitants, or, better still, be the native appellation.

3. Influence and Spread of Ideas

Stokes extended the exploratory work of Phillip Parker King and updated the work of earlier navigators and explorers, such as James Cook and Matthew Flinders. His contribution to the social and intellectual environment on board the *Beagle* is not fully documented. However, as the young Charles Darwin formed friendships with the officers Philip Gidley King, John Wickham and Stokes, some interchange of ideas must have taken place. The diligence and the enthusiasm shown by the young surveyors for their work, the discipline of the service, and the rigour and precision demanded by FitzRoy, created an atmosphere conducive to science.

The effects of Stokes' discoveries and surveys included the opening up of Endeavour Strait (off Cape York, the northernmost point of Queensland) for large vessels. The accurate charting of the Houtman Abrolhos, the Montebello Islands, and Bass Strait from the Victorian coast to Tasmania, and the southern part of the New Zealand coast, were significant in rendering passages and inlets safe for navigation. The addition of more observation points around Australia, each of which was fixed for latitude and longitude, where magnetic variation was measured, where the times of high water at full and changing moons were recorded, and the direction and speed of the tidal streams assessed, not only added to the existing knowledge of the country, but has provided a basis for present and future data analysis. The advice Stokes gave to various colonial governments on the location of lighthouses was another public service. Much of Stokes' work retained its value into the twentieth century. For example, the survey of Cockburn Sound in Western Australia was described by the historian Battye in 1912 as 'a work of extreme value even to the present day', and he wrote 'Much valuable information was gathered by the indefatigable efforts of Lieutenant Stokes, who with extreme methodical care collated it all for information of future parties' (Battye, 1912, p. 24). On Garden Island in Cockburn Sound, there is a road named after Stokes on the Royal Australian Navy base, HMAS *Stirling*.

Stokes' diplomatic handling of the subject of the location of Port Grey shows his ability to make objective conclusions based upon observation, yet without adversely criticizing the inaccuracy of the work of 'brother explorer' Captain George Grey, by then governor of South Australia. The object of the survey was to confirm the location of a harbour with a fertile hinterland, discovered by Grey in 1839. In a map of the area drawn by London cartographer John Arrowsmith, based on Grey's surveys, this harbour had been depicted some twelve miles south of Champion Bay. Stokes was familiar with that section of the Western Australian coast, having charted it in 1840, and was concerned that the London-based Western Australian Land Company was about to

> induce a large number of individuals, including women and children, by false, or at least exaggerated representations, to sever the ties of kindred and friendship, and become voluntary exiles to a far country, in search of a new and more prosperous home; whilst in lieu of the promised streams and fertile plains nothing in real awaited them but sterility – the certain loss of property, and the imminent risk of their lives. (Stokes, 1846, Vol. 2, p. 390)

The survey confirmed Stokes' impressions of the area, and proved Grey to be mistaken in his location of Champion Bay and his assessment of it as a harbour, surrounded by fertile country. Stokes expressed his criticism of Grey's geographical survey abilities in the most tactful terms:

> his description of the country refers to another portion; and it is only justice to him to state, that considering he was travelling for his life, and the great hardships he endured, it is surprising how the information collected was obtained. (Stokes, 1846, Vol. 2, p. 393)

With credit to Grey as a geographer, the correspondence between Stokes and Grey on the subject was made public in the Swan River colony's leading newspaper, the *Inquirer and Commercial News*, in May 1842. This exchange is an example of a dialogue between two scientific geographers, which in other circumstances would have been conducted under the auspices of the Royal Geographical Society.

The legacy of Stokes may be seen in the number of places named by him, which echo the events met by, and the officers of, the *Beagle*. Stokes is commemorated in Australia by Stokes Inlet and Stokes National Park in Western Australia, Stokes Range in the Northern Territory, and Stokes Point on King Island.

Stokes left a biological legacy in Bass Strait, and his comments on the matter are of great interest to island biogeographers concerned about the introduction of alien organisms. During the Bass Strait survey, he released

> about a dozen rabbits for the benefit of any unfortunate voyagers who might be thrown hungry ashore in this locality [East Cove, Deal Island]. During the few days that we were there they appeared to thrive very well, and I have no doubt that if not disturbed the island will soon be overrun with them, there being no wallabies to offer molestation. (Stokes, 1846, Vol. 2, p. 424)

Another island in Bass Strait had been populated by dogs left by sealers; and Rabbit Island, near Wilson's Promontory on the Victoria coast, was so called by the *Beagle*'s surveyors because of the abundance of rabbits 'multiplied from a single pair turned loose by a praiseworthy sealer six years before' (Stokes, 1846, Vol. 2, p. 426). These comments provide firm dates and other useful evidence for researchers investigating the history and effects of exotic introductions and biological invasions.

The link between Stokes and the administration and practice of hydrography may be seen from the appointments of two Hydrographers. Both had served under Stokes' command on the *Acheron*'s New Zealand survey in 1848–51. Upon the sudden death of the Hydrographer, Rear Admiral Washington (a founding member of the Royal Geographical Society) in 1863, Stokes was a potential successor to the position of Hydrographer. Although he was not appointed, perhaps through the lack of family connections to secure such a position, he had some indirect influence on the following twenty years of Admiralty hydrography. Two younger officers on the survey of the New Zealand coast were to play leading roles in scientific research in oceanography and magnetism. George Richards, Admiralty Hydrographer from 1864 to 1874, was Stokes' senior assistant and second in command, and Frederick Evans, Hydrographer from 1875 to 1884, had been master on the same survey. During Richards' period of office, there was a great rise of interest in the marine environment, especially deep-sea currents, the topography of the ocean bed, and the biota of the benthic zone. As Hydrographer, he contributed to the foundation of modern oceanographic science by appointing and equipping the *Challenger* expedition of 1873–6. Richards' successor, Evans, had made a vital contribution to navigation through solving the problem created by the effect of iron hulls and armour plating on the magnetic compass. His work on global magnetic variations earned him a Fellowship of the Royal Society. That Evans rose to the office of Hydrographer from master reflects his scientific acumen. In view of the calibre of the three surveyors on the *Acheron* on the New Zealand expedition, it is not surprising that the charts were still in use a century after they were made.

The degree of influence that Stokes had on Richards and Evans cannot be known for certain. However, as he was their commander in the early years of their surveying careers, it may be concluded that Stokes imparted some of his enthusiasm for geography and scientific research. Twelve of Stokes' charts were still in use up to the Second World War, and he has been described by a Royal Australian Navy hydrographer, Commander Ingleton, in 1944 as 'one of the most scientific of the early marine surveyors'.

Bibliography and Sources

1. OBITUARIES AND REFERENCES ON JOHN LORT STOKES

'Admiral John Lort Stokes', *Proceedings of the Royal Geographical Society and Monthly Record of Geography*, Vol. vii, No. 10 (1885), 673.

'Admiral John Lort Stokes', *The Times* (London), 13 June (1885), 12.

Battye, C., 'H.M.S. "Beagle", 1838–1841 – Minor Explorations, 1837–1840', in Battye, J.S. (ed.) *The Cyclopedia of Western Australia*, Hussey & Gillingham, London, 1912, 24–7.

Blewitt, M., *Surveys of the Seas: A Brief History of British Hydrography*, Macgibbon & Kee, London, 1957.

Bolton, G.C., 'Stokes, John Lort (1812–1882)', in *Australian Dictionary of Biography*, Vol. 2, 1788–1850, Melbourne University Press, Carlton, Victoria, 1966.

Hordern, M., *Mariners are Warned! John Lort Stokes and H.M.S.* Beagle *in Australia 1837–1843*, Melbourne University Press, Carlton, 1989.

Keynes, R.D. (ed.) *The Beagle Record*, Cambridge University Press, Cambridge, 1979.

May, W.E., *A History of Marine Navigation*, G.T. Foulis & Co., Henley-on-Thames, 1973.

Powell, A., *John Stokes and the Men of the* Beagle *– Discoverers of Port Darwin*, Occasional Papers no. 1, Library Services of the Northern Territory, Darwin, 1986.

Ritchie, G.S., *The Admiralty Chart. British Naval Hydrography in the Nineteenth Century*, Hollis & Carter, London, 1967.

2. WORKS BY JOHN LORT STOKES

1846 *Discoveries in Australia With an Account of the Coasts and Rivers Explored and Surveyed During the Voyage of H.M.S. Beagle, in the Years 1837–38–39–40–41–42–43*, T. & W. Boone, London.

1851 'A survey on the southern part of the Middle Island of New Zealand', *Journal of the Royal Geographical Society*, Vol. xxi.

1856 'On steam communication with the southern colonies (Australia and the Cape of Good Hope)', *Journal of the Royal Geographical Society*, Vol. xxvi.

Numerous charts were drafted by Stokes, and were published by the Hydrographic Office of the Admiralty and by the London map-maker John Arrowsmith.

3. ARCHIVAL SOURCES

Correspondence and reports by Stokes are held in: the Dyfed Archives, Haverfordwest, Wales; the Hydrographic Department (formerly Office), Taunton; the National Maritime Museum, Greenwich; and the Royal Geographical Society, London. One report, on the examination of Port Grey, is in the collection of *Exploration Diaries* held by the Battye Library, Perth, Western Australia. Reports and additional references to Stokes, and the expeditions in which he took part, may be found in the *Journal and Proceedings of the Royal Geographical Society* and the *Nautical Magazine*.

Marion Hercock is a postgraduate student in geography at the University of Western Australia.

Chronology

1811	Born 1 August at Scotchwell, near Haverfordwest, Wales
1824	Entered the Royal Navy in the *Prince Regent* as a first class volunteer
1825	Transferred to the brig *Beagle* as a midshipman; took part in the expedition led by Captain Phillip Parker King to survey the coast of South America
1831	Appointed mate and assistant surveyor on the *Beagle* for the South American survey under Commander Robert FitzRoy; formed friendship with Charles Darwin
1837	Commissioned as lieutenant; reappointed assistant surveyor on the *Beagle* for survey of Australian waters
1841	Appointed commander of the *Beagle*; married to Fanny Marlay
1843	Paid off the *Beagle* as commander upon return to Portsmouth
1845	Elected a Fellow of the Royal Geographical Society
1846	Published *Discoveries in Australia* by order of the Admiralty; advanced to post rank
1847	Appointed captain in command of paddle-wheel steamer HMS *Acheron* to survey New Zealand; death of Fanny Stokes
1848–51	Surveyed New Zealand and the resurveyed parts of the New South Wales coast
1854	Married to Louisa French Garratt (née Partridge), widow of H.J. Garratt
1856–7	Served on the council of the Royal Geographical Society
1860–3	Surveyed south coast of Devon
1864	Promoted to rear-admiral
1871	Promoted to vice-admiral
1876	Promoted to full admiral on the retired list
1878	Awarded the Flag Officer's Greenwich Hospital Pension
1885	Died at Scotchwell, 11 June

David Thompson

1770–1857

No portrait of
David Thompson
is known to be extant

C. Ian Jackson

For two centuries before the era of railway-building began in the 1870s, European exploration of the Canadian west was almost entirely an integral part of the fur trade. By the end of the first decade of the nineteenth century, the North West Company had created a chain of trading posts that extended almost to the Pacific Ocean; later in the century, the posts of the Hudson's Bay Company reached Fort Yukon in Russian America. The rivers and lakes that formed the highways of this trade were traversed by innumerable Indians and voyageurs, and a much smaller number of explorers, of whom Alexander Mackenzie is deservedly the most famous. It was, however, David Thompson who, almost single-handedly, surveyed the routes in use by the early nineteenth century, and produced the first reliable map, 123½ inch × 78 inch (314 × 198 cm), at a scale of about fifteen miles to the inch (1 : 950,400), of the vast area between 45° and 60° north, 84° and 124° west. When Thompson arrived at Fort Astoria, at the mouth of the Columbia, in July 1811, Gabriel Franchère noted accurately that

> This gentleman travelled as a geographer rather than as a fur-trader. During his stay of 7 or 8 days with us he had the opportunity of taking several observations, being in possession of a good sextant, and it seemed to me that he kept a regular journal. (Lamb, p. 88)

The lengthy entry on David Thompson in the *Dictionary of Canadian Biography* closes by describing him as 'one of Canada's best-known and loved historical figures' (Vol. 8, p. 883). This is scarcely consistent with Belyea's more recent remark that '"the man Thompson himself" is exceptionally elusive ... Almost nothing is known of Thompson apart from the documents he himself produced' (Belyea, 1994, p. xvii).

On one level, these apparently contradictory statements can be reconciled by recognizing that Thompson's achievements as explorer, surveyor and geographer are sufficiently outstanding that the fragmentary knowledge of the man behind these achievements is of secondary importance. This is, however, an over-simplification, since some of the main events in his career have become, for historians if not for geographers, a source of major, ongoing and sometimes acrimonious controversy. The fact that several of the key documents relevant to these disagreements either are themselves contradictory or appear to have been permanently lost or destroyed does nothing to diminish the debate.

Geographers cannot ignore these controversies, but Thompson's contribution to geography is barely affected by them. In his field career he travelled on foot, on horseback or in a canoe for more than 50,000 miles.

> He was the first white man to travel the full length of that 'mystery river' of the Pacific Northwest, the Columbia ... Canada and the world owes to this distinguished man the first reasonably accurate map of the Canadian West, embodying results of his surveys from 1792 to 1812. He surveyed the Canada–United States boundary line from Maine [an exaggeration; Thompson's boundary survey had its eastern origin in Ontario/New York, where the 45th parallel of latitude reaches the St Lawrence River.] to the Northwest Angle of the Lake of the Woods and prepared the appropriate maps. He was the first white man to canoe around the shoreline of Lake Superior for the purposes of survey and the first white man to fix the true source of the Mississippi and Missouri rivers. (Thomson, vol. 1, p. 212)

1. Education, Life and Work

David Thompson was born in England, of poor Welsh parents newly arrived in London. Two years later, a month after the birth of a younger brother, his father died. Thompson was educated, from the ages of seven to fourteen, at the Grey Coat Hospital in Westminster, then a charity school for boys and now a secondary school for girls.

He appears to have shown an aptitude for mathematics which, he said later, pointed him towards the Royal Navy. However, at the time when his education was completed, the Royal Navy was in one of its periods of reduced activity, between the American and French Revolutions, and 'my lot fell to be engaged to the Hudson's Bay Company' (Glover, 1962, p. xiv). In view of his later achievements, it is tempting to see the company's request 'to procure a Scholar who had a mathematical education' (Glover, 1962, p. 19) as a prescient recognition of the need for a surveyor and cartographer. Some in the company undoubtedly felt that way, but others did not. The last quarter of the eighteenth century was one when the company was making some effort to meet competition by establishing posts in the interior, after a century in which its traders had been able to remain at their posts on Hudson Bay and wait for Indian middlemen to bring the furs to them. Those who favoured expansion saw the need for survey and mapping, and Thompson was not the first apprentice with relevant skills to be recruited from the Grey Coat Hospital (Tyrrell, 1934, pp. 64–7). However, those who wished to maintain the old ways tried to ensure that the mathematical skills of their apprentice 'clerks' were focused on the arithmetic of the fur trade. David Thompson's letter of introduction from London to Churchill expressed this ambivalence:

> David Thomson [sic], our Apprentice for seven years who has been instructed in the Mathematicks and in the method of determining the situation of places by lunar observations: We recommend him therefore to the attention of our Chief that he may

> be kept from the common men and employed in the writings, accounts, & Warehouse Duty, and occasionally making Observations, so that he may by degrees be made capable of business & become useful in our Service. (Quoted in Tyrrell, 1934, p. 88)

After spending his first winter in company service at Churchill Factory, David Thompson was transferred to York Factory in September 1785: his journey with two Indian messengers on foot along the shore of Hudson Bay was his introduction to travel through the wilderness of Rupert's Land. A year later he was one of two Baymen sent to establish the third inland post of the company, South Branch House, on the South Saskatchewan River south of the present Prince Albert, Sask. He remained in the interior until 1791, travelling as far west as the foothills of the Rockies, where he spent the winter of 1787–8. During this first major expedition to the interior, the company apparently did not expect him to do any survey, and he had none of the necessary instruments with him.

Nevertheless, a series of events took place during these years that was crucial in shaping his character and career. He broke a leg at Manchester House in December 1788. The leg could not be set properly, and Thompson was brought down the Saskatchewan River to Cumberland House in the spring of 1789. There he remained for another year. The longer-term significance of this prolonged recovery was that during the winter of 1789–90 he 'regained [his] mathematical education' (Glover, 1962, p. 55) and in particular learned the principles of surveying and related astronomy from Philip Turnor. In doing so, 'By too much attention to calculations in the night with no other light than a small candle my right eye became so much inflamed that I lost its sight' (Glover, 1962, p. 55). The following spring, with Thompson's leg still not fully healed, Turnor chose Peter Fidler as his assistant in exploring and surveying the Athabasca country. Some commentators have seen Thompson's inevitable disappointment as continuing into a lifelong jealousy of Fidler. The evidence on this is, however, very thin.

The other defining event took place two years earlier, also at Cumberland House. As vividly recalled by Thompson more than sixty years later, during the summer of 1787 he was

> sitting at a small table with the chequer board before me, when the devil sat down opposite to me, his features and color were those of a Spaniard, he had two short black horns on his forehead which pointed forwards; his head and body down to his waist (I saw no more) was covered with a black curling hair, his countenance mild and grave; we began playing [draughts], played several games and he lost every game, kept his temper but looked more grave; at length he got up or rather disappeared. My eyes were open it was broad daylight, I looked around, all was silence and solitude, was it a dream or was it reality? I could not decide. Young and thoughtless as I was, it made a deep impression on my mind. I made no vow but took a resolution from that very hour never to play a game of chance, or skill or anything that had the appearance of them and I kept it. (Glover, 1962, p. 43–4).

Even Glover, who was critical of the image that emerged in the first half of the twentieth century of 'Thompson the seemingly faultless Sunday School hero', agreed that he 'possessed all the more commonplace Puritan virtues' (Glover, 1962, p. xii), something that could not be said of most fur traders of the period.

In 1790, with the latitude and longitude of Cumberland House determined, Thompson made a round trip to York Factory, surveying the usual route with a sextant borrowed from Turnor. Back at York Factory in 1791, his unpaid apprenticeship completed, he was now hired on a standard three-year contract, and the company in London also welcomed his competence in surveying: 'Every information that can tend to form a good Survey and Map of the Country Inland will always be particularly acceptable to us' (quoted by Glover, 1962, p. xxiv). Thompson had brought with him to Churchill in 1784 the 'Hadley's quadrant and Robertson's elements of navigation in two volumes' that were given to each scholar on leaving the Grey Coat Hospital, but these had remained at Churchill when Thompson left there in 1785. Now familiar with more advanced methods, Thompson requested the company in London to provide survey instruments instead of the suit of clothes that traditionally marked the end of an apprenticeship.

From 1792 to 1797, Thompson was primarily involved in unsuccessful efforts by the company to find a practicable water route to Lake Athabasca and so enable it to compete with the traders from Montreal, who were already exploiting the rich fur resources of the Northwest.

Although in 1794 David Thompson achieved the promotion to surveyor that he hoped for, succeeding Turnor, in 1797 he left the Hudson's Bay Company to join the competing North West Company. The reasons for this are debated by historians. One possible explanation is that the Hudson's Bay Company seemed to be entering a period of reduced capacity and retrenchment, due in large part to the war with France. In the previous decade, Thompson had joined the company because opportunities in the navy had been limited; now the situation was reversed, as both the army and navy were recruiting those who might have joined the fur trade. Thompson had in fact been nominated for a further promotion, to become 'Master to the Northward', but this position would have given him considerable responsibility for management of the fur trade, and consequently diminished the opportunities for exploration and survey.

The North West Company certainly made full use of Thompson's abilities. At Grand Portage on Lake Superior in the summer of 1797, he was commissioned by William McGillivray, Alexander Mackenzie and other Partners

> to learn the true positions of [the Company's] Trading Houses, in respect to each other; and how situated with regard to the forty ninth degree of Latitude North ... And wherever I could mark the line of the 49th parallel of Latitude [I was told] to do so, especially on the Red River. Also, if possible to extend my Surveys to the Missisourie River, visit the villages of the ancient agricultural Natives who dwelt there; enquire for fossil bones of large animals, and any monuments, if any, that might throw light on the ancient state of the unknown countries I had to travel over and examine. (Glover, 1962, pp. 131–2)

For several years, survey was his primary activity, with fur-trading mainly an incidental pursuit when winter travel was impossible. Fur-trading gradually became more important again, but between 1797 and 1812 there was only one year (1803) when he was not almost constantly on the move when the climate permitted. His travels took him as far as Lesser Slave Lake in the north-west (1799), almost to Churchill on Hudson Bay in the north-east (1804) and to the mouth of the Columbia River in 1812. In one fairly typical year (1800), for example, he left the North West Company's post at Fort George (approximately 53°N, 111°W) on horseback in the early spring, travelling up the North Saskatchewan River to Rocky Mountain House (52°N, 115°W). He left there on 5 May, this time by canoe, and was back at Fort George on 12 May. A week later he set off again to Grand Portage (48°N, 90°E). He then travelled back over the 61 portages between Grand Portage and Lake Winnipeg (Harris and Geoffrey, 1987, Plate 63) and up the Saskatchewan River to Rocky Mountain House. From 5 to 23 October and 17 November to 3 December he was again on horseback, exploring southwards from Rocky Mountain House along the eastern flanks of the Rockies (see Belyea, 1994, pp. 3–20).

Meanwhile David Thompson had contracted a 'country marriage' with a Métis at Île-à-la Crosse on 10 June 1799. Charlotte Small was the daughter of a Northwester, Patrick Small, and his Indian (probably Ojibwa) country wife. Unlike Patrick Small and many other fur traders of the period, David Thompson did not leave his country family behind when he left the west. The marriage was formalized in Montreal in 1812, and Charlotte, after bearing seven daughters and six sons, outlived her husband by only three months (died 1857).

In 1800, and again the following year, Thompson was continuing the North West Company's search for a practicable fur trade route through the Rockies to the Pacific. The Northwesters, based in Montreal, believed themselves at a disadvantage as compared to the Hudson's Bay Company, so far as access to tidewater was concerned. Montreal was much further from the fur areas of the early nineteenth century than the forts along Hudson Bay. Alexander Mackenzie had found in 1789 that the river which bears his name led to the Arctic Ocean, not the Pacific, and the route that he took to the Pacific four years later was not practicable for furs and trade goods.

The problem that David Thompson and the North West Company faced was not simply that of finding a practicable route through the mountains. A human barrier existed in the form of the Piegan (Blackfoot) Indians of the western plains. As described by Thompson:

> The Peegans ... are the most powerful of the western and northern plains, and by right of conquest have their west boundary to the foot of the Rocky Mountains, southward to the north branches of the Missisourie, eastward for about three hundred miles from the Mountains and northward to the upper part of the Saskatchewan. (Glover, 1962, p. 252)

The Piegans were strongly opposed to direct trade between the fur companies and Indians living west of the Rockies.

Thompson's first attempt to cross the mountains in June 1801 failed because of an incompetent guide and unusually heavy rains flooding the gorges of the Ram River. Thompson felt that 'had the Waters been two Feet perpendicular lower, we might perhaps have succeeded' (Belyea, 1994, p. 33).

The North West Company then temporarily lost interest in the venture. Thompson became a partner in the company in 1804, and spent the next two years in the 'Muskrat Country' of the Canadian Shield, between the Churchill and Nelson Rivers. As in his previous employment with the Hudson's Bay Company, it appeared that fur-trading was being given a higher priority than exploration and survey, and according to Nicks, 'His career as a surveyor might have come to an end with his rotation on furlough in 1808, so dissatisfied was he by 1804–5 at not being able to pursue his interests in exploring and surveying' (Nicks, 1985, p. 881).

That furlough, however, had to be postponed for several years, as events caused the North West Company to renew its desire to establish itself beyond the western mountains. The overland expedition of Lewis and Clark reached the Pacific coast at the mouth of the Columbia river in 1805–6, raising the threat of subsequent American competition for western furs. By 29 October 1806 Thompson was back at Rocky Mountain House, and the following year, with his wife and family, he crossed the height of land in Howse Pass and spent the winter of 1807–8 trading furs on the upper Columbia River. In 1808 he brought these furs as far east as Rainy Lake (93°W), and then returned over the mountains. In 1809 and early in 1810 he completed an exploration of the upper Columbia and Pend d'Oreille Rivers that showed that they and the Howse Pass were not a practical 'route to the sea' for the fur trade. He spent the winter of 1809–10 at a new trading post, Saleesh House, near what is now the community of Thompson Falls in north-west Montana.

David Thompson's actions in 1810 and 1811 have generated more historical dispute even than his abrupt decision to leave the Hudson's Bay Company in 1797. Another factor becomes significant here, and that is national sovereignty. To several Canadian historians, notably Arthur Morton in the 1930s and Richard Glover in the 1960s, David Thompson is a principal reason why the Canadian boundary from the Rockies westward lies on the 49th parallel, instead of much further south on the latitude of the Columbia River.

In the spring of 1810, Thompson again crossed the Rockies, heading east with the product of his fur-trading. He anticipated continuing on to Montreal for his postponed furlough, but was met at Rainy Lake with instructions from the North West Company that led to a further postponement. These instructions have not survived, but they caused Thompson to return once more to Rocky Mountain House and to cross the Rockies yet again, this time on his way to the mouth of the Columbia River.

To Thompson's critics, these lost instructions had set in motion a race to the mouth of the Columbia, to establish Britain's claim to the area, ahead of the known similar ambitions of Jacob Astor and his Pacific Fur Company. '[T]he Nor'Westers could not have intended him to run his race to the sea in such a manner as to lose it ... Yet Thompson did so run his race as to lose it' (Glover, 1962, pp. lii–liii).

A more generous view of the issue is that there was no such race. Instead, Thompson's journey can be seen as serving a dual purpose: putting into effect a cooperation between the two companies, while ensuring that the North West Company's interests were protected. It is now known that, at Fort William in July 1810, the wintering partners had approved a one-third interest in Astor's enterprise, though this was ultimately never ratified (Bridgewater, 1949, pp. 52–4).

> Thompson may simply have been instructed to survey a feasible route across the mountains to the mouth of the Columbia. Once there, he could represent the NWC, ensuring that the Astorians did not cut into the trade already established by the NWC in the interior. The few surviving documents tend to support the latter interpretation. (Nicks, 1985, p. 881)

These, however, are just the bare bones of a very complex controversy concerning Thompson's motives and actions. The geographer is primarily interested in Thompson's actual journey, in the course of which he completed the survey of the Columbia River all the way to its mouth.

Thompson's original intention was to take his usual route through the Howse Pass, but it seemed that the Piegans were by now determined 'to block any further trading from the interior across the mountains' (Nicks, 1985, p. 881). Instead, he set off northwards at the end of October 1810, and, in snow and temperatures that reached as low as −36°F (−38°C), reached the height of land at 5700 feet (1740 m) above sea level on 10 January 1811. (The altitude is that given by Morse (1969, p. 107); Tyrrell in 1916 stated that it was 6025 feet, and Thompson's *Narrative* of the journey, written forty years after the event, quotes Simpson that it was 'computed to be eleven thousand feet'.) In his *Narrative*, Thompson stated that on this day 'my object was to be at the Pacific Ocean before the month of August'. Thompson's critics see this as an after-the-fact rationalization (Glover, 1962, pp. li–lii, 322).

The object was achieved. On 15 July 1811, 'we came to the House of Mr Astors

Company' (Belyea, 1994, p. 155, and see below) at Astoria, a house on which construction had begun only in April. It was then that Gabriel Franchère described Thompson as a geographer rather than a fur trader.

After completing his survey of the Columbia River, Thompson crossed the Rockies twice more that year, collecting supplies from Henry's House on the upper Athabasca River before wintering for the last time in western America at Saleesh House.

After further exploration and survey, David Thompson and his family left Saleesh House on 13 March 1812, crossed the mountains for the last time via the Athabasca Pass, and reached Fort William on Lake Superior on 12 July. A month later he travelled eastwards again, resurveying the northern shore of the lake, and eventually arrived in Montreal.

For the next three years the surveyor turned cartographer. The North West Company granted Thompson a salary, and a full share of the profits of the company for three years, in which time he was to convert his journal observations into maps for the company. The main product was the huge map covering the area between latitudes 45°N and 60°N and longitudes 84°W and 124°W. The original version became the property of the North West Company and has since disappeared. A revised version was prepared soon afterwards and one surviving copy is in the Ontario Archives (Hopwood, 1971, p. 327).

Formally retired from the North West Company, in 1815 Thompson moved to a farm at Williamstown, Upper Canada. In January 1817, at the age of 46, he was appointed astronomer and surveyor on the British team for the boundary survey created by the Treaty of Ghent. The first stage of the survey covered the portion of the international boundary between the St Lawrence River and Sault Ste Marie and was completed in 1822. According to Nicks (1985, p. 882), Thompson and the British commissioner Anthony Barclay 'were subjected to a good deal of personal criticism' regarding the agreed boundary, since 'Upper Canadian opinion widely regarded it as a sell-out of Canadian interests'. A different opinion was taken by James White nearly a century later:

> Thompson had made surveys of the greater portion of Western Canada and had found that the deepest channel was usually much nearer the *north* side of rivers. When the survey was completed the United States commissioner contended for the deepest channel, but Barclay insisted on the letter of the treaty. The British Admiralty desired that Wolfe Island, opposite Kingston, be obtained. If it passed to the United States, fortifications could be erected on it that would threaten the British navy yard and forts. Barclay was successful in obtaining Wolfe Island in exchange for Grand Island, above Niagara Falls, and Barnhart and other islands, near Cornwall ...
>
> Reviewing the division of the islands in the River St Lawrence, there can be no doubt that Great Britain fared well. (White, 1913, pp. 829, 836)

In the second stage, from Lake Superior to the Lake of the Woods (1822–7), Thompson was the only official astronomer, enjoying the confidence of both the British and US commissioners, although the interpretation of the delimited boundary remained in dispute in two places (west of Lake Superior and the Northwest Angle) and ultimately had to be resolved through the Ashburton–Webster Treaty of 1842 (Gentilcore, 1993).

Completion of work on the international boundary coincided with the beginning of the decline in Thompson's financial situation, which continued until his death. One reason for this was the bankruptcy of the firm that had acted as agents for the North West Company; another is usually given as Thompson's inadequacy in managing his financial affairs. Glover remarked that 'what staggers one about his hardships is his apparent failure to inspire the most commonplace family affection among his children' (Glover, 1962, p. lxv), who among them might have been expected to prevent his descent into penury.

His financial problems caused him to continue to seek work as a surveyor into his seventies and led, in 1844, to him embarking on the work for which, with his map of western America, he is mainly remembered. This, following the usage when it was eventually published in 1916, is normally referred to as David Thompson's *Narrative*. Hopwood, however, preferred the term *Travels*, on the grounds that this is how Thompson himself referred to the work as it was being written. The work did not produce the modest income that Thompson had hoped for, partly because he constantly revised it and was unable to produce a volume ready for publication. Another major reason was that in 1848 his remaining eye began to fail. Thomson quotes a medical text by Henry Howard, published in 1850, recording the following case:

> David Thompson, Surveyor and astronomer, led to my surgery by his daughter on February 24, 1848. Claimed to have been blind in right eye since February 1789. Claims he woke up one morning so blind he could not tell daylight. (Thomson, Vol. 1, p. 211)

Thompson apparently temporarily regained sight in his left eye, and continued his daily journal until 28 February 1851.

David Thompson died in obscurity in Montreal in February 1857. 'His reputation in geographical matters might have been largely lost but for the laudable and zealous salvage work of JB Tyrrell of the Geological Survey of Canada' (Thomson, Vol. 1, p. 212).

2. Scientific Ideas and Geographical Thought

Perhaps the most amazing characteristic of David Thompson was that 'the greatest land geographer of his day in British America and the maker of the first comprehensive map of the Canadian West' left school in 1784 at the age of 14, and from then until 1812 was almost always on the frontier of the fur trade, never returning to Britain and never going east of Lake Superior. There were other great travellers and surveyors of the period in the fur trade. However, Philip Turnor arrived in the area from Britain at the age of 26, already trained. Alexander Mackenzie went to London during the winter of 1791–2 to learn the techniques and acquire the instruments that he needed if he was 'to return to find a passage through those western mountains' (quoted by Thomson, Vol. 1, p. 202). Peter Pond was a major explorer and adequate cartographer, but his locations were sometimes flawed, especially in regard to the Mackenzie River and Great Slave Lake.

David Thompson's surveys were carried out in a period when three major improvements in technique were transforming the sciences of navigation and survey. The sextant was replacing the quadrant, the *Nautical Almanac* provided basic astronomical information from 1767 onwards, and the chronometer was providing an easy solution to the problem of determining longitude. Thompson benefited from the first two of these, but not the last. The chronometer developed so successfully by Benjamin Harrison during the eighteenth century was intended mainly for maritime use, and even at sea it took many decades for such expensive devices to be widely adopted. Thompson had received a quadrant on leaving the

Grey Coat Hospital, and had there learned the rudiments of its use; his introduction to the sextant came during the winter of 1789–90 at Cumberland House, thanks to Philip Turnor. In the opinion of the historian of Canadian mapping and survey, Turnor's 'greatest contribution to Canada's development was in training David Thompson and Peter Fidler in the sciences of surveying, mapping and the making of field notes' (Thomson, Vol. 1, p. 204).

It appears that all the surveying instruments used by Thompson up to 1812 were acquired during his employment with the Hudson's Bay Company (Smyth, 1981, p. 3). They included a sextant, magnifying glass (to read the vernier scale on the sextant), artificial horizon, magnetic compass, telescope, thermometers and watches. In addition, he had drawing instruments, the indispensable volumes of the annual *Nautical Almanac* and its associated *Requisite Tables*, and Robertson's textbook *The Elements of Navigation*.

In both his Hudson's Bay Company and North West Company employment, surveying was an adjunct to the fur trade, not merely in terms of its utility but also in the way that it was carried out. The key surveying elements were (a) determination of latitude and longitude at specific points (often along a river that Thompson the fur trader was travelling); (b) keeping a record, with the aid of a magnetic compass, of changes in direction during travel; and (c) 'dead reckoning' (i.e. estimation), on the basis of time taken (Belyea, 1994, p. 183), of distances travelled, the latter subsequently corrected in terms of astronomical calculations.

What this meant in terms of Thompson's daily journals, on which his great maps were subsequently based, can be illustrated by quoting, in their entirety, from the entries recording his arrival at Astoria in July 1811. This was undeniably a major event in his life, yet the journal entries are typically factual in character. The form in which the surveying details are recorded is very similar to that of his teacher Philip Turnor (see Tyrrell, 1934). Belyea has noted that the astronomical calculations and 'course notations' occupied between a quarter and a third of each of Thompson's Columbia journals.

> **July 15th [1811] Monday** A very fine day, somewhat cloudy. Staid 'till 6:25 Am shaving & arranging ourselves, when we set off, Co [course] N33W 1M [mile] N65W 2M N78W 1M S70W 1M S60W ½M SW 1½M the Fog all along prevents me seeing well S34W 2M S22E ½M So ½M S50W 1/5 W^{t} 1/6 SW ½ + ⅔M S50W 2M + 1M N68W 1M + 1½ M to Pt Tongue, but as the Wind was blowing from the Sea very hard, we made a portage of ab^{t} 100 Y^{ds} over this Tongue and again embarked, Co to the Ho [House] S50W 1½M. At 1 Pm, thank God for our safe arrival, we came to the House of Mr Astors Company – Messrs McDougall, Stuart & Stuart – who received me in the most polite Manner, & here we hope to stay for a few days to refresh ourselves.
> **July 16th [1811] Tuesday** A fine day. Obs^{d} for Lat^{de} Long^{de} & Time Lat^{de} 46°:13′:56″N by Acc^{t} Long^{de} 123°:48¼′W. (Belyea, 1994, p. 155)

Current atlas coordinates for Astoria are 46° 12′ N, 123° 50′ W (*Times*) or 46° 11′ N, 123° 49′ W (*Rand McNally*).

Of the various instruments carried, it was Thompson's watches that were least reliable. According to Smyth:

> Thompson did not have the use of chronometers during his western surveys. He relied on the accuracy of locally-set watches and carried with him at least two watches at all times. If not moved after setting, these would have kept reasonably accurate time for short periods up to 24 hours. (Smyth, 1981, p. 7).

It was therefore necessary for Thompson to determine local time by astronomical means on almost a daily basis, weather permitting.

A complete set of observations would therefore entail:

(a) determination of latitude, normally based on the altitude of the sun at noon, corrected for light refraction through measurement of air temperature;
(b) determination of local time, based on knowledge of latitude, normally using the rising or setting sun;
(c) determination of longitude, as the difference between local time and time at the Greenwich meridian, normally using the method of lunar distances and correcting for parallax and refraction.

Smyth has noted that 'A single observation for longitude was followed by three to four hours of mathematical calculations, employing the aid of *Nautical Almanac, Requisite Tables* and, likely, Robertson's *Elements of Navigation* as well' (Smyth, 1981, p. 13). And the 'single observation' of longitude involved seven steps in recording time, temperature and astronomical altitudes.

David Thompson normally spent several months each winter at a specific location, such as Rocky Mountain House. In such circumstances, there was adequate opportunity for repeated observations of both latitude and longitude, and it is not surprising that his determinations are highly accurate. Sebert (1981, pp. 412–13) has noted that some of the differences may be attributable to errors in the *Nautical Almanacs* of the period, especially in regard to lunar distances, rather than to Thompson's mistakes in observation or calculation. The means of 10 listings of latitude and 19 of longitude for Rocky Mountain House differ by about one mile in latitude and five miles in longitude from the modern determination (Smyth, 1981, p. 15).

On the trail, it was, of course, a different matter. Observations had to be taken whenever possible, recognizing the restrictions on observing imposed by weather conditions as well as by the rigorous schedule of canoe and horseback journeys that several times spanned half a continent and back in the course of a single summer. What is truly amazing is the frequency of such observations, maintained over two decades of exploration and travel. In 1811, for example, between the beginning of May and mid-September, David Thompson made 35 determinations of latitude and 34 determinations of longitude on his journey down the Columbia River to Astoria and back again (Belyea, 1994, pp. 314–16). Much earlier, in 1798, Alexander Mackenzie had met Thompson at Sault Ste Marie:

> Upon my report to him of the surveys I had made and the number of astronomical Observations for Latitude, Longitude and Variation of the Compass, he was pleased to say I had performed more in ten months than he expected could be done in two years. (Glover, 1962, p. 219)

Hopwood (1988, p. 207) has described 'this 4,000 mile circle in ten months' as 'surely one of the greatest feats of surveying in all history'.

Given such dedication, Thompson's abrupt transfer from the Hudson's Bay Company to the North West Company in 1797, and his contemplated departure from the fur trade a decade later, no longer seem so surprising or inexplicable. Very few individuals could have maintained such an obvious commitment to the innumerable details of survey for such a long period; for someone who could, like David Thompson, a life devoted only to the fur trade must have seemed impossibly mundane. A further conclusion seems inescapable. Throughout those two decades of travel, exploration and survey, Thompson must surely have contemplated the creation of the vast map of western America that he produced in the two years after leaving the field (Hopwood, 1988, p. 205). Otherwise, the immense detail that he had collected, expressed at least as much in the 'course' detail as in the latitude and longitude determinations, would have been quite excessive and even pointless.

Widely different evaluations have been made of the accuracy of Thompson's survey, beginning with the anonymous but evidently disgruntled note in the Selkirk Papers dated 1 November 1816:

> [The observations] lately published all rest on the authority of a Mr David Thompson of the NW Co whose survey on a large scale is hanging up in the great Hall of this Fort [Fort William]: nothing can be more rediculously incorrect than every part of it which we have had an opportunity of scrutinizing ... But the neat drawing, the minuteness & apparent care bestowed on his plans may well have deceived people into the belief of their accuracy, so long as there were no other surveys to compare with his. (Quoted in Belyea, 1994), p. 188)

Other less biased critics have been generous in their assessments. Tyrrell, whose own travels through the areas that Thompson surveyed made him a legendary figure, commented that 'While my instruments may have been better than his, [Thompson's] surveys and observations were invariably found to have an accuracy that left little or nothing to be desired' (Tyrrell, 1916, p. xxxii, quoted by Belyea, 1994, p. 188). About his later survey and map of the international boundary west of Lake Superior, Henry Youle Hind commented in 1858: 'His map ... is an admirable piece of work. We recognized every portage as we came to it ... although Thompson's survey was made in 1826' (quoted by Thomson, Vol. 1, p. 215.) A similar judgement was made much more recently about Thompson's survey of the Pend Oreille valley in Washington state: 'Thompson's records are remarkable for their geographical accuracy in all cases where he notes details of his journeys, and in such instances his progress can be followed almost step by step' (Smith, quoted by Sebert, 1981, p. 405).

There is, however, an unresolved problem concerning the inaccuracy of some of Thompson's determinations of latitude, since it is in principle much easier to get accurate astronomical observations of latitude than of longitude. Stewart (1936) prepared a table showing errors in the location of 13 points along the Saskatchewan and North Saskatchewan Rivers, which led him to conclude that:

> It is a given fact, then, which we have to accept, that the latitudes and longitudes given in Thompson's notes cannot, in general, be accepted as defining accurately the position of old trading posts or other points of interest. (Quoted by Sebert, 1981, p. 413).

That criticism, however, should probably be restricted to determinations of latitude and longitude made in the course of Thompson's actual travels. Where he spent a period of time at a specific site, even if only several days, there was the opportunity for multiple observations, and consequently reliable determinations. Even for the locations cited by Stewart, Sebert considered that, although 'some of the latitudes are surprisingly inaccurate, ... it is difficult to imagine anyone working with the instruments and under the conditions that Thompson worked being able to consistently read more accurate longitudes' (Sebert, 1981, p. 413).

One possible explanation for the 'surprisingly inaccurate' latitudes (six of the thirteen locations are misplaced by three miles or more) is that on occasion Thompson appears to have extrapolated or interpolated latitude, based on the course travelled between sites where direct latitude determinations were possible (see Thompson's journal entry for 11 July 1811 and Belyea's note on this entry, Belyea, 1994, pp. 153 and 274).

A further comment is that Thompson's surveys were all two-dimensional. The man who surveyed so much of the western cordillera seldom mentioned altitude in his journals, and, as noted above, later quoted the height of land in the Athabasca Pass at 11,000 feet above sea-level, whereas it is not much more than half that altitude. It was recognized in the eighteenth century that the mercury barometer could be used as an altimeter (Wilford, 1981, p. 98), but this was scarcely a practical method for Thompson's type of travel, and the aneroid barometer was not introduced until 1843. Another alternative was trigonometric levelling, using the sextant as a substitute for a theodolite. Thompson was aware of this technique, and used it on at least one occasion. On 2 April 1808, he calculated the height of 'Mount Nelson' as 7223 feet (2202 m) above the level of Kootenay Lake, which is at 1745 feet (532 m) above sea-level, although Thompson had no means of knowing the lake's altitude. (Tyrrell, reproduced in Glover (1962, p. xcii), appears to have used an incorrect lake altitude.) Efforts of this type, requiring a long, measured baseline, were feasible only when Thompson spent considerable time at a site. On his 1813 map his indication of mountain ranges, using hachures, shading and stylized summits, is always generalized rather than specific.

3. Influence and Spread of Ideas

There is a significant and mildly amusing division of opinion on whether David Thompson's cartography or his writing is the more important legacy. For Thomson, the map of the Canadian west 'may, in the long future, represent David Thompson's chief claim to enduring fame' (Thomson, Vol. 1, p. 205). Smyth concurs: 'Thompson's prominent position in Canadian history (his remarkable *Narrative* notwithstanding) rests largely on the accuracy of his comprehensive surveying and mapping of a then largely uncharted wilderness' (Smyth, 1981, p. 1).

A very different view is taken by Glover and Hopwood. Glover distorts Franchère's description of Thompson at Astoria: 'as Franchère says, with the appearance of a mere geographer' (Glover, 1962, p. l), and elsewhere comments that

> His best work in his last years was the writing of his *Narrative*. This is indeed his major work. His mapping was, of course, an outstanding achievement and can in nowise be belittled; but it was certain that North America would some day be mapped, and any competent, adequately financed surveyor, or team of surveyors, could have done it. Only Thompson, on the other hand, could have produced the *Narrative*. (Glover, 1962, p. lxv)

The irony of this statement, in which Glover denies that Thompson's mapping can be belittled, and then proceeds immediately to do just that, is best answered by Warkentin's remark that Thompson's great map

> was the base for many maps of Western Canada until the Dominion Land Surveyors began their work in the 1870's. Even in the twentieth century, Thompson's map served as a base map for some areas lying beyond the districts laid out by the land surveyors. (Warkentin, 1964, p. 67)

Hopwood, though fully recognizing Thompson's surveying and cartographic achievements, ultimately takes the same view as Glover: 'The *Travels* [i.e. *Narrative*] in the long run prove his greatest memorial. Even maps as great as Thompson's maps are superseded by better maps' (Hopwood, 1971, 329).

To some extent, the difference of opinion is immaterial: why should it be necessary to choose between the map and the *Narrative*? Both are nowadays regarded as geographical and historical works of the first order. Diminution of the map on the grounds that sooner or later others would map the area, and map it more accurately, seems an odd judgement for a historian; one might equally disparage the *Principia* on the grounds that other natural philosophers would ultimately have defined Isaac Newton's laws of physics. Similarly, those who have found fault with the *Narrative* on the grounds either that it was written long after the events it records, when its author was nearly eighty years of age, or that it tends on occasion to be selective and self-serving where Thompson's own actions are concerned, have not significantly reduced its unique authority as a description of a vast area and its people (especially aboriginal people) at a time when white settlement was virtually non-existent.

A more important reason for regarding this 'map versus *Narrative*' debate as unproductive is that it tends to ignore the third leg of the stool that is Thompson's legacy. Without Thompson's journals, the in-the-field notebooks in which he recorded all the data he collected, neither map nor *Narrative* would have been possible. These journals, most of which are in the Ontario Provincial Archives, have largely remained unpublished and are seldom even quoted. Belyea was able to claim in 1994 that her volume 'is the first major edition of Thompson's journals' (Belyea, 1994, p. xxiv). As the short extracts quoted earlier indicate, the spare, factual style of the journals does not make for enthralling, or even easy, reading. Also, in Belyea's view:

> Apparently historians have sensed that Thompson's text does not yield easily to historical methods and aims. Despite the immediacy of the genre – the writer tells of his own experiences as they occur – Thompson's journals resist attempts to 'go behind' the textual surface in order to reconstruct the author and his times ... Thompson's journals are an *aide-mémoire*, not a communication in the usual sense ... The hundreds of daily entries form a pattern at once intimate and impersonal, candid and purely professional. (Belyea, 1994, pp. ix–x).

As such, of course, they are the source books not merely for the maps and the *Narrative*, but for contemporary and future studies of the geography of western Canada and the Columbia Basin in the late eighteenth and early nineteenth centuries, and of a wider area still; Hopwood notes that in addition to 22 years of journal-keeping in the west, Thompson maintained the practice for another 38 years, until blindness intervened at the age of 80.

It is often noted that the North West Company was remarkably wise when, having attracted David Thompson from its great rival, the Hudson's Bay Company, it enabled him to spend several years focusing on the task of exploration and survey. But those years were only a relatively small part of both his period as a Northwester and his longer career as explorer, surveyor and cartographer. Such work was clearly his main interest, but for most of the time that he spent west of the 84th meridian it had to be accommodated to the demands of the fur trade. From his fur-trade years, and to a lesser extent from his work on the boundary surveys, he left a legacy of a detailed day-by-day record of observations, that he himself developed into outstanding cartography and geographical literature. It is sad that, whereas his cartography was widely appreciated and used for most of the nineteenth century, his

Narrative remained unpublished, and therefore unused, until well into the twentieth century. But that in no way diminishes his own achievement as a geographer, summarized as follows by a professional geographer of our own day:

> For the Narrative, Thompson occasionally described the country through a brief itinerary of his route, in the style of Sir Alexander Mackenzie. But more frequently he prepared a geographical synthesis of a region, with the topography accurately depicted, the resources carefully assessed, and the human geography described ...
>
> One of Thompson's greatest achievements in the Narrative is his perceptive description of the Great Plains. Only a fully disciplined mind, capable of bringing order out of fragmentary detail, could have delineated a geographical region so simply and yet so masterfully ...
>
> Thompson had a sound conception of the geographical regions of the Canadian West. He is one of the first men whose writing conveys to the reader the variations in regional character that are found within the Western Interior of Canada, and the essential qualities of each region. (Warkentin, 1964, p. 93)

Bibliography and Sources

1. SELECTED BIOGRAPHICAL SOURCES

Belyea, B., *Columbia Journals, David Thompson*, McGill-Queen's University Press, Montreal and Kingston, 1994.

Bond, R., *The Original Northwester: David Thompson and the Native Tribes of North America*, Spokane House Enterprises, Nine Mile Falls, Washington, c. 1972.

Glover, R. (ed.), *David Thompson's Narrative 1784–1812*, The Champlain Society, Toronto, 1962.

Hopwood, V.G. (ed.), *David Thompson: Travels in Western North America*, Macmillan of Canada, Toronto, c. 1971.

Hopwood, V.G., 'David Thompson and his maps', in Farrell, B. and Desbarats, A. (eds), *Explorations in the History of Canadian Mapping: A Collection of Essays*, Association of Canadian Map Libraries and Archives, Ottawa, 1988, 205–10.

Morton, A.S., *David Thompson*, Toronto, Ryerson Press, 1930.

Morton, A.S., 'The North West Company's Columbian enterprise and David Thompson', *Canadian Historical Review*, Vol. 17 (1936), 266–88.

Nicks, J., 'David Thompson', in *Dictionary of Canadian Biography*, Vol. 8, University of Toronto Press, Toronto, 1985, 878–84. Includes a lengthy list of sources, including archive material, up to 1980.

Smith, J.K., *David Thompson, Fur Trader, Explorer, Geographer*, Oxford University Press, Toronto, 1971.

Tyrrell, J.B. (ed.), *David Thompson's Narrative of his Explorations in Western America 1784–1812*, The Champlain Society, Toronto, 1916.

2. GEOGRAPHICAL MATERIAL PRODUCED BY DAVID THOMPSON

A. Manuscript

All but one of Thompson's surviving journals are in the Ontario Provincial Archives in Toronto. Normal access is by microfilm (MS 25, 8 reels) assisted by a finding aid (F443). One journal, dealing with the abortive attempt to cross the Rockies in 1801, is in Vancouver Public Library. The original of the *Narrative* is in the University of Toronto Library, except for one chapter (IIA, pp. 37–56 in Glover, 1962) which is in the Ontario Provincial Archives. Other manuscript material is in the National Archives of Canada (Ottawa), and in London in the Public Record Office and the Library of the Royal Commonwealth Society.

Hopwood (1988) listed the following principal maps drawn by Thompson, though he did not cite the sources on which he relies:

(a) A map prepared for the North West Company during his survey of the locations of its posts in 1797–8.

(b) A map, now lost, prepared during his winter at Fort George 1799–1800. 'This map ... is probably the basis of the maps that Alexander Mackenzie published in 1801 in the account of his journeys to the Arctic and Pacific Oceans.'

(c) The original 'Map of the North-West Territory of the Province of Canada' prepared in 1813, that hung in the North West Company's Great Hall at Fort William on Lake Superior. It has since been lost or destroyed. This map, however, appears to have been a source of maps published by Arrowsmith and others for about 20 years, since both it and they include the non-existent 'Caledonia River' near what is now Seattle.

(d) The 'Map of the North-West Territory of the Province of Canada', revised by Thompson in 1814 to exclude (on the evidence from Stuart's explorations in 1812) the spurious 'Caledonia River'. This map is now on permanent exhibition in the Ontario Provincial Archives.

(e) Thompson's maps of his boundary survey from the St Lawrence to the Northwest Angle, published by the US government in 1898.

B. Editions (in addition to the editions by Belyea, Glover, Hopwood and Tyrrell cited above)

Burpee, L.J. (ed.), 'Some letters of David Thompson', *Canadian Historical Review*, Vol. 4 (1923), 105–26.

Coues, E. (ed.), *New Light on the Early History of the Greater Northwest: The Manuscript Journals of Alexander Henry and of David Thompson*, 3 vols, 1897. Reprinted (2 vols) by Ross and Haines, Minneapolis, 1965.

Elliott, T.C. (ed.), 'Journal of David Thompson', *Oregon Historical Quarterly*, Vol. 15 (1914), 39–63, 104–25, 216.

Elliott, T.C., 'David Thompson's journeys in the Spokane country', *Washington Historical Quarterly*, Vol. 8 (1917), 11–16, 103–6, 169–73, 284–7; Vol. 10 (1919), 17–20.

Elliott, T.C., 'David Thompson's journeys in Idaho', *Washington Historical Quarterly*, Vol. 11 (1920), 97–103.

Elliott, T.C., 'David Thompson and beginnings in Idaho', *Oregon Historical Quarterly*, Vol. 21 (1920), 49–61.

Elliott, T.C., 'The discovery of the source of the Columbia River', *Oregon Historical Quarterly*, Vol. 26 (1925), 23–49.

Elliott, T.C., 'David Thompson's journeys in the Pend Oreille country', *Washington Historical Quarterly*, Vol. 23 (1932), 18–24, 88–93, 173–6.

Howay, F.W. (ed.), 'David Thompson's account of his first attempt to cross the Rockies', *Queen's Quarterly*, Vol. 40 (1933), 333–56.

Leverette, C.E., *David Thompson's Journal of the International Boundary Survey, 1817–1827 [and] western Lake Erie, August–September, 1819*, Killaly Press, London, Ontario, 1974. This is a cyclostyled set of extracts from Thompson's journals. One copy is in the Robarts Library at the University of Toronto.

Morse, E.W., *Fur Trade Canoe Routes of Canada – Then and Now*, Queen's Printer, Ottawa, 1969.

Schoefield, E.O.S. (ed.), *Report of the Provincial Archives of British Columbia (1912)*, King's Printer, Victoria, 1913.

White, M.C. (ed.), *David Thompson's Journals Relating to Montana and Adjacent Regions, 1808–1812*, Montana State University Press, Missoula, 1950.

Wood, W.R. (ed.), 'David Thompson at the Mandan-Hidatsa villages, 1797–1798', *Ethnohistory*, Vol. 24 (1977), 329–42.

Wood, W.R. and Thomas, D.T. (eds), *Early Fur Trade on the Northern Plains: Canadian Traders Among the Mandan and Hidatsa Indians 1738–1818: The Narratives of John Macdonell, David Thompson, François-Antoine Laroque and Charles McKenzie*, University of Oklahoma Press, Norman, 1985.

3. OTHER SOURCES CITED IN THIS BIOBIBLIOGRAPHY

Bridgwater, D.W., 'John Jacob Aster relative to his settlement on Columbia River', *Yale University Library Gazette*, Vol. 24, no. 2 (1949), 47–69. This collection of documents, from the Astor Papers in the Beinecke Rare Book & Manuscript Library at Yale University, was apparently unknown to Glover. It contains probably the best surviving evidence that there was no 'race to the sea' in 1810–11, and its significance is discussed by Lamb (1969, pp. 10–14).

Gentilcore, R.L. (ed.), *Historical Atlas of Canada*, Vol. II: *The Land Transformed*, University of Toronto Press, Toronto, 1993, Plate 21.

Harris, R.C. and Geoffrey J.M. (eds), *Historical Atlas of Canada*, Vol. I: *From the Beginning to 1800*, University of Toronto Press, Toronto, 1987, Plates 61–3, 66–7.

Lamb, W.K. (ed.), *[Gabriel Franchère's] Journal of a Voyage on the North West Coast of North America during the Years 1811, 1812, 1813 and 1814*, The Champlain Society, Toronto, 1969.

Rich, E.E., *The History of the Hudson's Bay Company 1670–1870*, Hudson's Bay Record Society, London, 2 vols, 1959. Rich's history includes the competition between the Hudson's Bay, XY and North West Companies, and therefore covers the whole period of Thompson's travels in the west, mainly in Chapters 5–7 and 10–11 of Volume 2.

Ronda, J.P., *Astoria and Empire*, University of Nebraska Press, Lincoln, c.1990.

Sebert, L.M., 'David Thompson's determination of longitude in western Canada', *The Canadian Surveyor*, Vol. 35, No. 4 (1981), 405–14.

Smyth, D., 'David Thompson's surveying instruments and methods in the northwest 1790–1812', *Cartographica*, Vol. 18, No. 4 (1981), 1–17.

Stewart, W.M., 'David Thompson's surveys in the northwest', *Canadian Historical Review*, Vol. 17 (1936), 289–303.

Thomson, D.W., *Men and Meridians*, 3 vols, Queen's Printer, Ottawa, 1966–9.

Tyrrell, J.B. (ed.), *Journals of Samuel Hearne and Philip Turnor*, The Champlain Society, Toronto, 1934.

Warkentin, J. (ed.), *The Western Interior of Canada, A Record of Geographical Discovery 1612–1917*, McLelland and Stewart, Toronto, 1964.

White, J., 'Boundary disputes and treaties', pp. 751–958 in Shortt, A. and Doughty, A.G. (eds), *Canada and Its Provinces, A History of the Canadian People and Their Institutions by One Hundred Associates*, Vol. 8, Toronto, Publishers' Association of Canada, 1913, 751–958.

Wilford, J.N., *The Mapmakers*, Knopf, New York, 1981.

C. Ian Jackson lives in Hamden, Connecticut. He is a director of Chreod Ltd, Ottawa, Ontario and an Associate Fellow of Timothy Dwight College, Yale University.

Chronology

J.B. Tyrrell compiled, in considerable detail, 'David Thompson's itinerary in north-western America, 1785–1812.' This was reproduced in Glover (1962, pp. lxxiii–cii).

1770	Born in London (Westminster) 30 April
1772	Brother John born 25 January; father died 28 February
1777–84	Grey Coat Hospital School, Westminster
1784	Apprenticed to Hudson's Bay Company for seven years; sailed from London 29 May, arrived Churchill Factory early September
1785	Transferred, by land along shoreline, to York Factory, September
1786–9	Travelling across the prairies, broke leg at Manchester House 23 December 1788 and unable to walk until August 1789
1789–90	Wintered at Cumberland House and trained as surveyor by Philip Turnor; became permanently blind in right eye; brother John entered Hudson's Bay Company service 1789 (until 1797)
1791	Returned to York Factory. Hired on three-year contract when apprenticeship ended
1792–7	Surveying and fur-trading for Hudson's Bay Company, mainly in the 'Musk Rat Country' between the Nelson and Churchill Rivers; contract renewed for three years in 1794; nominated in 1796 to succeed Ross as 'Master to the Northward' in 1797
1797	Quit Hudson's Bay Company at Bedford House on completion of

	contract and 13th anniversary of apprenticeship (21 May); joined North West Company a few days later
1797–1812	Surveying and fur-trading on behalf of the North West Company throughout what is now western Canada and adjacent areas of the USA (see Tyrrell's 'Itinerary')
1799	'Country marriage' to Charlotte Small on 10 June at Île-à-la Crosse (where she was born 1 September 1785)
1800	Daughter Fanny born 10 June at Rocky Mountain House
1804	Son Samuel born 5 March, Peace River Forks; became a partner in the North West Company, July
1806	Daughter Emma born March, Reed Lake House (died 22 February 1814)
1808	Son John born 25 August, Boggy Hall (died 11 January 1814)
1811	Son Joshua born 28 March, Fort Augustus
1812	Left Saleesh House (in present Montana) 13 March, via Athabasca Pass, Cumberland House and Fort William, arriving in Montreal mid-August; marriage to Charlotte formalized 30 October; settled in Terrebonne, north of Montreal, October
1812–15	Retained as full partner by North West Company to prepare maps based on his western surveys
1813	Son Henry born 30 July, Terrebonne (died 23 October 1855)
1815	Daughter Charlotte born 7 July, Terrebonne; family moved to Williamstown, Glengarry Co. (present Ontario)
1817	Appointed astronomer and surveyor for the Boundary Commission created by the Treaty of Ghent, January; daughter Elizabeth born 25 April, Williamstown
1817–21	Surveying and mapping international boundary under 6th Article of Treaty of Ghent (St Lawrence River to Sault Ste Marie)
1819	Son William born 9 November, Williamstown
1820	Appointed justice of the peace, Glengarry County
1822	Son Thomas born 10 July, Williamstown
1822–7	Surveying and mapping international boundary under 7th Article of Treaty of Ghent (Sault Ste Marie to the Lake of the Woods)
1824	Son George born 13 July, Williamstown
1827	Daughter Mary born 2 April, Williamstown
1829	Daughter Eliza born 4 March, Williamstown
1825–33	Increasing financial difficulties leading to near-bankruptcy in 1833
1833–44	Carried out a variety of surveying tasks in the present Ontario and Quebec, but financial difficulties continued, with loss of Williamstown home
1844–51	Wrote his *Narrative*, initially intended as a money-raising venture, but

	no part of it was published in his lifetime; lost sight in remaining eye in 1851
1857	Died 10 February, Longueuil, Lower Canada (present Quebec). Buried in Mount Royal Cemetery, Montreal

Charles Warren Thornthwaite

1899–1963

John Russ Mather
and Marie Sanderson

Phillips Studio, Philadelphia

C. Warren Thornthwaite was possessed of that rare spark of genius that allowed him to rise from a poor rural farm background in central Michigan to become a university professor and a world-renowned climatologist and geographer. Unquestionably, he was the most outstanding American climatologist of the twentieth century.

Thornthwaite originally achieved fame in geographical and climatic fields as the originator of a new, rational classification of climate in 1948. A more permanent contribution to world research was his concept of potential evapotranspiration and his water balance model, which formed the basis of his climatic classification. The water balance model compares water need or potential evapotranspiration with water supply or precipitation. It has continued to be used by scientists in many disciplines for almost fifty years. Its simplicity and ease of verification have made the model a useful tool for hydrologists, forest and agricultural scientists, and engineers. The model has been used to estimate surface runoff in areas for which no runoff data exist and to extend the runoff record back in time. It provides data on soil moisture and evapotranspiration that are very useful for agriculture and forest scientists. The provision of quantitative estimates of water deficiency has similarly proved of practical use to crop scientists and irrigation engineers.

Perhaps Thornthwaite himself did not visualize the many future uses of his water budget technique. As a college professor, his original goal was to develop a new climatic classification; later, as a consulting climatologist, he began to realize that his novel concept of potential evapotranspiration and his water budget model had practical uses in many fields of science. He could thus be called the father of applied climatology. Through his innovative research and writings, and his contacts with climatologists world-wide, Thornthwaite profoundly influenced the development and direction of the modern field of climatology.

1. Education, Life and Work

Charles Warren Thornthwaite was born on 7 March 1899 on a farm near Pinconning in central Michigan, USA, the first of four children (two boys and two girls). His father was a farmer and harness-maker. Hard work on the farm, clearing land, planting, harvesting – often by hand – milking, and gathering wood for fires were all part of his early lot. His parents soon learned that the only way to escape the backbreaking farm life was through education, an understanding that directed Warren, as he preferred to be called, for the rest of his life.

There was no high school in Pinconning, the nearest town, so Warren was sent to Mount Pleasant to live after he graduated from grade school. He managed to obtain a job as janitor of the local Methodist church, which allowed him to live in a room in the basement. He graduated from high school in 1918 and was inducted into the Army Cadet Corps. The First World War ended within a few months and the corps was disbanded. He quickly enrolled in Central Michigan Normal School (now called Central Michigan University) in Mount Pleasant to earn credentials to become a schoolteacher. There he met his future wife, Denzil, and also a future colleague in climatology and geography, John Leighly. At Central Michigan, Warren and Leighly took geography from R D Calkins, who was instrumental in helping Leighly obtain an Assistantship to work with Carl Sauer and the department head W.H. Hobbs at the University of Michigan. Sauer left Michigan for the University of California at Berkeley in 1923 and persuaded Leighly to join him as an assistant and to work for a PhD.

Thornthwaite graduated from Central Michigan in 1922 (his favourite courses were mathematics and music rather than geography) and obtained a job teaching high school in Owosso, Michigan. During the summers of 1923 and 1924 he did take summer courses at the University of Michigan. In 1924, Warren was persuaded by John Leighly to return to full-time college life and to work towards a PhD in geography at the University of California at Berkeley under Carl Sauer. Berkeley was rapidly developing as a leading centre for geographical thought, and while Sauer's main interest at that time was historical geography, he did teach an introductory course in physical geography. More important, however, was the fact that many of the geographical scholars of the day were brought by Sauer to Berkeley as visiting lecturers, so that Thornthwaite was exposed to the ideas of such eminent thinkers as Albrecht Penck, William Morris Davis and others. It was an atmosphere charged with excitement and the never-ending joys of intellectual discovery.

Money was a problem for Thornthwaite, who was now married and so, in 1925, he began to work part-time with the Kentucky Geological Survey in Frankfurt, Kentucky. Sauer was also interested in urban geography at this time and he suggested that Thornthwaite undertake research on the city of Louisville, Kentucky, for his doctoral dissertation, which he did. While the dissertation was never published, two American urban geographers recently reviewed his dissertation and concluded that Thornthwaite was as far ahead of his geographical colleagues in his understanding of urban problems and in analytical techniques as he was later in climatology. They wondered just how differently the field of urban geography in the USA might have developed if Thornthwaite had followed this early interest in urban geography rather than climatology. It is quite likely that Thornthwaite never understood the significance of his unique contributions to the field of urban geography.

Thornthwaite was awarded his doctorate from Berkeley in 1930, although he had already obtained a position as an assistant professor of geography at the University of Oklahoma in 1927. The Department of Geology was renamed the Department of Geology and Geography in 1921, and by 1923 the geography courses were listed separately and a major programme in geography was created. The Department of Geography was transferred to the College of Arts and Science from the School of Mines in 1927, and Thornthwaite joined the department as its second faculty member at that time. The first two undergraduate degrees in geography were issued by the department in 1926. The first Master's degrees were granted in 1930. Each semester, until he left the University of Oklahoma, Thornthwaite taught four or five courses, an incredibly heavy load for someone finishing his doctoral programme and undertaking a research programme. For example, in Autumn 1927, he taught world regional, physical, economic geography, geography of North America, and seminars in geography, while in spring 1928 he added general geography, cartography, industrial geography, and geography of Asia. In the next five years, he added courses on urban geography, geography of Oklahoma, geography of Africa and Australia, and principles of human geography. One might note that in that grand array of courses, there was not one called climatology, for Thornthwaite had not yet specialized in what was to become his life's work.

Thornthwaite's first publications were issued by the University of California Press and involved base maps of California (1926) and of Eurasia and Africa (1927) and a paper on a cylindrical equal-area map projection (1927). His first paper on weather and climate came in 1929, dealing with the polar front and its role in the interpretation and prediction of Oklahoma weather. It is clear that climate was becoming a growing interest at this time, for only two years later he published his first major article (in the *Geographical Review* of the American Geographical Society) entitled 'The climates of North America according to a new classification' (1931). This was followed in 1933 with 'The climates of the earth', also published in the *Geographical Review*, and Thornthwaite's career in climatology was fairly well established. Nearly all of his papers and monographs after that time dealt with some aspect of climatic classification, the climatic water budget, agricultural climatology, and aspects of applied climatology.

While living in Oklahoma, Thornthwaite, as an eclectic geographer and student of Carl Sauer, became interested in the economic and cultural development of the Great Plains region. This led ultimately to an invitation from the University of Pennsylvania's Study of Population Distribution to join them in a serious study of the Great Plains. He left Oklahoma in 1934 to undertake this new assignment. Several publications (*Internal Migration in the United States*, 1934, and 'The Great Plains', 1936) resulted from his work on this project. In 1933, President Roosevelt had established the Soil Erosion Service, and in 1935 its name was changed to the Soil Conservation Service and it was moved to the Department of Agriculture. Because of his earlier papers on climate classification and his ongoing work on the Great Plains, Thornthwaite was asked by the vice-director of the service to join the work of the Soil Conservation Service as head of the Climatic and Physiographic Research Section, which he did. There was considerable interest in conservation work in those days and Thornthwaite lost no time in building up his staff and initiating research work at stations located across the country. By September 1935 he had permission to establish a climatic research centre in western Oklahoma. (The country was now in the midst of the Dust Bowl conditions of the mid-1930s, and so this was a high-priority research undertaking.) Climate and erosion work was also begun on the Polacca Wash in Arizona (with the advice and support of his old mentor, Carl Sauer). His section pushed the development of a large research centre at Coshocton, Ohio in the Muskingum Valley as well as in South Carolina.

Thornthwaite was busy writing articles and giving speeches during his decade with the Soil Conservation Service (1935–45). A paper entitled 'The significance of climatic studies in agricultural research' introduced his growing interest in agricultural climatology, and the application of climatic data in the solution of applied problems. Another paper described his growing interest in the problem of the year-to-year variability of climate.

Two major research papers appeared during these years with the Soil Conservation Service. The first summed up his pioneer work in measuring evapotranspiration from a grass field in Arlington, Virginia (now the site of the Pentagon) for a whole year using sensitive wind- and humidity-measuring instrumentation. This was an epochal accomplishment in view of the rather crude state of meteorological sensors at that time and could only be accomplished with careful design of each instrument and the use of a significant amount of manpower available at the time as a result of the lingering depression. Leighly read the report of this year-long experiment and wrote

> I'm already seeing in my imagination such installations as yours over corn fields and irrigated alfalfa fields and above the crowns of forests; and a new class of climatic data from everywhere, namely evaporation from the natural surface of the ground. It is a magnificent prospect. (Mather, 1995)

The second outstanding paper published during his Soil Conservation Service days was a joint article with his old colleague, John Leighly. His collaboration with Leighly was so successful that there was serious consideration of a joint book to be entitled 'Physical and Applied Climatology'. A favourable response was received from a publisher and the project was ready to get under way when US participation in the Second World War stopped further progress. It is of interest to remember that no similar book was available at the time and none appeared for nearly thirty years. This indicates how far Thornthwaite was ahead of his colleagues in recognizing what was needed if climatology was to take its place as a rigorous physical science rather than as merely a descriptive addition to geography or meteorology. Had such a book appeared at that time, it might have led to a rapid advance in the science of climatology, an advance that ultimately required three decades to achieve.

In 1943, the article, co-authored with Leighly, entitled 'Status and prospects of climatology', appeared in the *Scientific Monthly*. On discussing the development of climatology (as opposed to meteorology) in the USA, Thornthwaite and Leighly pointed out that the most pressing need was to develop the area of microclimatology which involved understanding the patterns of climatic distributions between the surface of the earth and the tree tops. Climate over a farm or a city was of more significance than was the world pattern of climate in solving practical problems of everyday life. Thornthwaite and Leighly went on to point out the lack of any place in the USA where adequate training in climatology and especially microclimatology could be obtained. They called for the creation of

> a new agency, an institute for climatologic research ... The institute should be affiliated with a first-rate university which includes an agricultural college and experiment station ... An exceedingly important task of the institute would be the training of climatologists ... on the graduate level. (Mather, 1995)

This recognition of the need for an institute for climatic research and training was published in 1943, and within four years, almost serendipitously, the pieces needed for its creation began to fall into place.

With the prosecution of the war effort, there was a decline in interest in agricultural research and the work of the Climatic and Physiographic Division began to wane. Thornthwaite contributed to the war effort by advising the Mexican government on irrigation requirements of sugar cane in order to expand production of that crop, and this led him into studies of scientific irrigation for a wide range of crops. Since the need for water was related to the amount of evapotranspiration from the plant and soil surfaces, he increased his efforts to find ways to evaluate vegetation demands for water. He designed soil-filled tanks buried flush with the ground and kept constantly moist to measure how much water the plants would use. He called the water use under these ideal soil moisture conditions 'potential evapotranspiration' to distinguish it from the daily water use of vegetation under varying soil moisture conditions and under a mixture of vegetation covers. The first real discussion of the concept of 'potential evapotranspiration' along with a formula to compute it from readily available climatic data appeared in a brief note in the *Transactions of the American Geophysical Union* in 1944.

It was in late 1945 that Thornthwaite was asked by John Seabrook, a Princeton-trained officer in Seabrook Farms (possibly the world's largest grower and packager of frozen garden vegetables at that time), to come to the farms in southern New Jersey to advise on the irrigation of crops and orchards. Seeing this as a new challenge in an area of interest since his childhood – to expand the role of scientific agriculture in order to ease the life of the farmer and to increase food production – he took leave of the Soil Conservation Service and spent the growing season of 1946 as a consultant to Seabrook Farms.

The challenges were far more exciting and interesting than he had ever imagined. His leave of absence from the Soil Conservation Service was extended for a second year and after that he terminated all ties with the government and struck out on his own as a private consultant – first to Seabrook Farms and shortly thereafter to other food processors. Supported in part by a large multi-year research contract from the US Air Force dealing with micrometeorology of the surface layer of the atmosphere as well as by his consulting fees from industry, he saw the opportunity to establish the institute for climatic research that he had proposed some four years previously. He created at Seabrook a Laboratory of Climatology associated with the Johns Hopkins University (to manage the government research contract) and located at Seabrook Farms (a first-class agricultural organization that would take the place of an experimental station as identified by Thornthwaite) and invited a number of graduate students to come to study climatology and to work for Masters degrees or doctorates at Johns Hopkins. The plan outlined in his 1943 paper was in place and operating under the name of the Johns Hopkins University Laboratory of Climatology. After six years, the Air Force terminated its micrometeorology contract and Johns Hopkins wanted the laboratory to move its operation to the Baltimore campus of the university. Thornthwaite had no desire to move back to a city area and to lose the opportunity to continue his research in the agricultural fields of southern New Jersey. A new university sponsor (Drexel Institute of Technology) was located to handle the growing number of research contracts that the scientists at the laboratory received.

Possibly the most significant of his publications appeared in 1948, when the *Geographical Review* published his paper on 'An approach toward a rational classification of climate'. This was the culmination of his work to devise a more rational classification based on potential evapotranspiration and the climatic moisture index. He felt that this new classification had incorporated the one aspect missing in earlier classifications, since it allowed him to compare plant water need with the climatic supply of water from precipitation. This provided a true understanding of the relative moistness or dryness of a climate – an understanding that could not be obtained from a comparison of precipitation with the actual evapotranspiration. The scales of potential evapotranspiration and the moisture index were not established arbitrarily or as a result of some vegetation or soils boundary, as was the case with earlier climatic classifications. Rather, the boundary of each climate type was identified on the basis of a rational progression of climatic data from one boundary to the next.

With the ability to compute the climatic demand for water (potential evapotranspiration) from readily available climatic data, it was now possible to evaluate a climatic water balance or budget as a day-by-day or month-by-month comparison of the water demand with the water supply. During the 1950s, much of the work at the laboratory involved applying the results from the climatic water budget to the solution of various environmental problems. For example, it was used to determine the flow in the Nile River and the monthly contributions from the White and Blue sources of the river, to determine the fate of Lake Maricaibo or the Mediterranean Sea if access to the ocean were to be blocked by dams, to determine irrigation schedules for shade-grown tobacco in Connecticut and cotton in Arkansas, to determine the ability of vehicles or humans to move over unpaved surfaces (soil tractionability), and to determine the recharge of moisture to the groundwater table under different types of land cover.

The Laboratory of Climatology became a Mecca for climatologists from all over the world, and there was hardly a day that a research scholar from some distant place was not in residence, lecturing, collaborating in research, or learning what was being done at the laboratory. This was what Thornthwaite envisioned as the role of the institute: research, teaching, and international collaboration to develop the discipline of climatology and to disseminate widely the knowledge gained. The German scholar, Rudolf Geiger, the father of micrometeorology, made only one trip to the USA, but spent more than two months teaching at the laboratory, where he found a stimulating intellectual atmosphere.

Consulting work expanded rapidly and so Thornthwaite was forced to establish a new group, C.W. Thornthwaite Associates, to handle the many requests. These included questions of scheduling irrigation for various farming enterprises, disposing of industrial waste water by spray irrigation in unused woodlands, and the manufacture and sale of a range of sensitive micrometeorological instruments to measure profiles of wind, temperature and dew point temperature as well as net radiation, soil heat flow and vertical wind movements. At one time, the laboratory and associates employed as many as eight PhDs along with more than 25 research technicians and support personnel on both research and consulting activities.

With the publication of the 1948 rational classification and the establishment of the Laboratory of Climatology, Thornthwaite became well known internationally. The World Meteorological Organization (WMO) was established in 1947 when it took over the functions of the long-established International Meteorological Organization (IMO). The WMO was an association of 79 member states of the United Nations and had the task of regulating world meteorology because, of all the sciences, it represented the one most dependent on international accord.

The first congress of the WMO was held in Paris in 1951, and eight technical commissions were established to help keep the WMO abreast of problems in each major part of the field. One of these was the Commission for Climatology (CCL). Thornthwaite was not present at that congress because he was not a representative of any member government. In spite of this, he was elected first president of CCL by its members. In his presidential address to the First Session of the Commission for Climatology in Washington in March 1953, he referred to his rather unusual election:

> I regard my election to the Presidency of this commission, in Paris two years ago without my knowledge or consent, as further evidence of a change in emphasis. I am not a meteorologist in the narrow sense, but a geographer turned climatologist ... Therefore, my election seems to me to have been a further indication that our commission has a mandate from WMO to do whatever can be done to promote the science of climatology.

To understand these comments it should be stated that the last president of the Climatology Commission (a dynamic meteorologist) felt that the task of the commission was to create a closer connection between climatology and dynamic meteorology.

The first session of the CCL, held in the State Department in Washington, lasted for two weeks and was attended by delegates from 19 member states along with representatives from other United Nations agencies, the International Union of Geodesy and Geophysics, and the International Geophysical Union. Thornthwaite suggested to the members that they were free to think about the development of climatology in both the scientific and applied fields independently of meteorology. He urged the creation of an international institute for climatic study and research. Clearly, he saw his Laboratory of Climatology as a model for such an Institute. The CCL set up subcommittees to work with the World Health Organization to study how meteorological data could be applied in the study of climate and health and with the Food and Agricultural Organization to apply climatic classification to the introduction of new food crops world-wide. The question of the arid zones of the world was raised, and it was felt that the climatic factor was not being properly weighted in the analysis of problems of arid regions. The CCL endorsed the production of homoclimatic maps of the arid zones as a first step in understanding the growing problems of arid zones.

Thornthwaite served a second term as president of the CCL until January 1957. During this second four-year term, the principal emphasis of the commission as far as Thornthwaite was concerned was the creation of a world climatic atlas. Thornthwaite felt that there was widespread interest in such an atlas among the technical commissions and regional associations, and so he proceeded to build support, especially with the Food and Agricultural Organization and UNESCO. Even without formal support, Thornthwaite started working on a few sheets for the proposed atlas – water budget factors for south-west Asia and for eastern North America – to serve as models for other meteorological services to follow. However, no other support was forthcoming at that time. Perhaps his concept of a world climate atlas, spiral-bound so that sheets could be changed as data were updated, was too grand for the member governments at that time.

Thornthwaite stepped down as president of the CCL at the end of his second term in early 1957. He was satisfied that climatology had taken rapid steps forward in developing independently from meteorology and that significant basic and applied research was well under way in the field.

Thornthwaite became a well-recognized 'world citizen' through his travels in connection with the WMO, UNESCO, and the CCL. In 1957, he participated in a

WMO-sponsored regional training seminar on the water balance and hydrologic forecasting in Belgrade, Yugoslavia. He then went on to a UNESCO meeting on arid zone problems in Karachi, Pakistan. Thornthwaite liked to point out that he crossed the Atlantic 15 times from west to east but only 14 times from east to west. Clearly, one of his many trips involved a continuation around the world so that he did not return by way of the Atlantic. Later in 1957, Thornthwaite attended a meeting of the UNESCO Arid Zone Committee in Canberra, Australia, and served as chair of the meeting. He subsequently lectured in Honolulu, Melbourne, Singapore, Bangkok, Karachi, Tehran and Geneva. His international reputation had reached its peak.

Thornthwaite received a number of prestigious awards for his work in climatology. Among the most important was the Cullum Geographical Medal of the American Geographical Society, given to those who had distinguished themselves by geographical discoveries or the advancement of geographical sciences. Such notable explorers as Robert Peary, Fridtjof Nansen, Robert Falcon Scott and Sir Ernest Shackleton were previous Cullum Medal winners. He was also selected as Honorary President of the Association of American Geographers from 1960 to 1961. In his presidential address at that group's annual meeting, he launched a fairly strong attack on the regional geography of the day:

> For the most part, the title 'geographer' means University Professor, the centre of activity is the classroom, and a 'regional specialist' is one who can organize and teach a regional course in a creditable fashion ... That growing numbers of geographers entertain different views as to the content and objectives of geography and as to the training required for productive work in the field is attested by the increasing emphasis on quantitative methods ... the student who has mastered any of the systematic branches of geography, and who has a solid foundation in physical science to back it up, makes a better geographer. (Thornthwaite, 'The task ahead', 1961, p. 347)

Recognizing Warren Thornthwaite's unique contributions to climatology and geography world-wide, Central Michigan University sought to honour possibly its most outstanding graduate and chose to award him an honorary doctorate in June 1963. Unfortunately, Thornthwaite had to receive this award *in absentia*; the ceremony took place just nine days before his death from cancer. His last few years had not been particularly happy ones, for while he was finally receiving recognition from the world for his achievements and contributions, his health was failing. After the death of his wife in 1962 his health problems became more severe and he spent much of the last six months of his life in hospital, where he died on 11 June 1963.

2. Scientific Ideas and Geographical Thought

Thornthwaite was an eclectic climatologist and geographer. His interests in both areas were widely focused, yet two central underlying themes permeated his writings. The first was based on his belief that a strong mathematical and scientific background was needed to do rigorous work in either field. He wrote:

> There has never been any doubt that to do acceptable work in any of the physical branches of geography such as geomorphology, climatology, hydrology, or pedology, mastery of mathematics and physics, together with an extensive knowledge of chemistry, are required. The present work of the regional scientists being published in

> the *Journal of Regional Science*, some of which is outstanding, brings into focus the need for genuine competence in mathematics in regional geography as well ... During recent years a few young geographers have demonstrated the power of modern mathematical and physical science by using them in the solution of problems in economic and cultural geography and in various population problems. (Thornthwaite, 'The task ahead', 1961, p. 348)

The second underlying theme in nearly all his research and writing was that of application. As a realistic and practical individual, he viewed all scientific endeavour as goal-oriented. He did not view science for science's sake, but rather science for some practical application and for the solution of pressing problems. Thus, even his work on climatic classification, which engaged him for more than thirty years of his life, was directed towards understanding the role of moisture in climate, for better assessment of the climatic potential of an area, and for improving the role of agriculture through understanding possible crop introductions into new areas, the role of drought in climate, and the distribution of natural vegetation. He became identified as the father of applied climatology, defined as the application of climatic data to the solution of operational problems. These two themes, while often unexpressed, were the basis of much of his research and writing.

Thornthwaite produced no books, since he had so many ideas to disseminate that time for book writing was never available. He did produce more than 100 journal articles and monographs which clearly set forth in simple terms his thoughts and ideas about the whole subject of climatology and especially the moisture factor in climate, agricultural climatology, and the future of climatology. Thornthwaite was not a complex writer. His thoughts were expressed in simple, short sentences. His close friend F. Kenneth Hare wrote about his writing style:

> Warren was a true genius, with a remarkable intuitive ability to identify how complexes work. But he was impatient of science, and very suspicious of formal mathematical logic. Much of what is now seen as the core of physical climatology, bioclimatology and ecosystem theory was clear in his mind, but not in his papers. He neither could (because of an unwillingness to submit to drill) nor would write an orthodox defence of his seminal ideas.

Organization of Thornthwaite's work in climatology and geography is difficult but five major themes seem to be found in his writings. These might be listed as climatic classification, the water budget, agricultural climatology, microclimatology and instrumentation, and finally the future of climatology.

The earlier part of his life focused largely on work on the first two themes: climatic classification and the water budget. As a Sauer-trained geographer and an acute observer of the environment, as well as an avid reader of the research works of others, he was fully aware of the problems that all earlier climatologists experienced when trying to express quantitatively the role of the moisture factor in climate. His concept of potential evapotranspiration provided a solution to many of the problems of climatic classification. Where annual precipitation (*P*) exceeded the annual climatic demand for water, potential evapotranspiration (*PE*), climates were moist and forests predominated. Where *P* was less than *PE*, deserts or semi-arid steppes existed. In those places where annual *P* essentially equalled annual *PE*, subhumid climates with grassland vegetation were found. The term 'subhumid climates' was new with Thornthwaite and solved one of the problems found with the more common Köppen classification. A drawback to the Thornthwaite classification was that it required four separate maps to depict the distribution of each of the factors of the classification, making it complex and unwieldy, and inconvenient for classroom display and teaching. While no world maps of the four elements of the classifications have appeared, many of Thornthwaite's papers provided the classification for individual countries: 'Problems in classification of climates' (1943); 'The moisture factor in climate' (1946); 'Climate and moisture conservation' (1947); 'The role of evapotranspiration in climate' (1951); 'The water balance in tropical climates' (1951); 'Grassland climates' (1952); 'The water balance in arid and semiarid climates' (1953); 'Climatic classifications in forestry' (1955); and 'The water balance' (1955).

Following this activity in developing and modifying the climatic water budget, Thornthwaite turned his attention to solving practical problems in agriculture and water resources. Papers marking this period in Thornthwaite's research development concerned both the applied climatology and agricultural climatology aspects of his interests. They included: 'Climate in relation to planting and irrigation of vegetable crops' (1952); 'Climate and scientific irrigation in New Jersey' (1952); 'Operations research in agriculture' (1952); 'Climate in relation to crops' (1954); 'Climatology and irrigation scheduling' (1955); 'The water budget and its use in irrigation' (1955); 'Estimating soil moisture and tractionability conditions for strategic planning' (1958); 'Investigations of the climate and hydrologic factors affecting the redistribution of strontium-90 in the soil' (1959); and 'Movement of radio-strontium in soils' (1960).

A fourth series of investigations centred on microclimatology as well as on instrumentation to measure profiles of weather elements near the ground. Thornthwaite was strongly influenced by Rudolph Geiger's famous work *Climate of the Layer of Air Near the Ground*, the first edition of which appeared in 1927, but he added to Geiger's monumental work the ability to improve meteorological sensors so they would truly record short-time-period changes in weather elements of importance. Although Thornthwaite was not trained in instrumentation, he had an innate understanding of what was needed in microclimatic sensors (wind, radiation and dew point sensors). This work was summed up in part in the following publications: 'Microclimatic studies in Oklahoma and Ohio' (1937); 'A dew point recorder for measuring atmospheric moisture' (1940); 'The chemical absorption hygrometer as a meteorological instrument' (1941); 'Note on the variation of wind with height in the layer near the ground' (1942); 'Final report, micrometeorology of the surface layer of the atmosphere' (1954); 'Microclimatic investigations at Point Barrow, Alaska' (1956, 1957, 1958); 'Modification of rural microclimates' (1956); 'Dew point apparatus' (1957); 'Modified SCS cup anemometers' (1957); 'Measurement of vertical winds in typical terrain' (1959); 'The measurement of vertical winds and momentum flux' (1961); and 'Vertical winds near the ground at Centerton, N.J.' (1961).

As he matured as a climatologist and geographer, Thornthwaite became more interested in the development of the field of climatology and in changing the direction of geography. He held strong views on how climatologists should be trained (in much the same way that meteorologists are trained – calculus, physics, statistics, biology, atmospheric dynamics, meteorology). But he saw a rigorous physical science background as desirable for geographers as well. Unfortunately, he had no background in computer programming, so he did not include that as one of his recommended areas for training, but if he were alive today, it certainly would be on his list. He was somewhat outspoken in his views and unfortunately alienated many of his colleagues, especially in geography, with his criticism of current developments in the field. Writings that touch on these sensitive subjects include:

'Status and prospects of climatology' (1943); 'Modern meteorology and its application to the agricultural regions of the United States' (1950); 'The task ahead in climatology' (1957); 'American geographers: a critical evaluation' (1961); 'The task ahead' (1961); and 'The geographer's role in climatology' (1962).

3. Influence and Spread of Ideas

Long-time friend and colleague F. Kenneth Hare probably summed up peer evaluation most successfully in his obituary:

> Warren Thornthwaite was not an easy man. He did not suffer fools gladly, nor did he take kindly to criticism. Much of his scientific writing does scant justice to the genius that inspired it. He never submitted to the rigor and discipline required by a high scientific style. Yet the brilliance and originality of his insight were unquestionable to those who knew him best ... Without doubt he was the most influential climatologist of his generation. (Hare, 1963, p. 597)

In contrast, his friend and colleague from college days, John Leighly, evaluated Thornthwaite's legacy quite differently:

> To judge from Thornthwaite's repeated exposition of his procedure for estimating the water balance of terrestrial surfaces ... he looked upon this as his most important achievement. Yet it was seriously questioned in his lifetime and may now be considered obsolete. (Leighly, 1964, p. 615)

The authors disagree with Leighly. Judging by the number of papers still being published which use the water budget and the myriad of applications of water budget information to the solution of practical problems such as leaching through the soil, recharge to the water table, stream flow, soil trafficability, and irrigation scheduling, it is clear that, far from being obsolete, the water budget is still being recognized as a most useful analytical tool in hydroclimatology.

The institute for climatic research and training envisioned by Thornthwaite more than 50 years ago still exists, but the changes that have occurred in the academic and research world have made the need for such an institute less important. Because climatological research and teaching are now being conducted at many universities, government agencies and private research organizations, there is less need for a single centre to supply the research and training that Thornthwaite envisioned in 1943.

In a survey of the *Science Citation Index* which lists the number of citations of the work of scientists, citations of Thornthwaite's works were found to average 53 each year from 1980 to 1984, 58 each year from 1985 to 1989, and in the last three-year period, from 1990 to 1992, 62 each year. Thornthwaite was fourteenth in the total number of citations, just ahead of his friend and mentor Carl Sauer. In a list of only physical geographers, Thornthwaite was sixth. In a survey of climatology books published over the past three decades, discussions of some aspects of Thornthwaite's work were found in 21 of 28 books. An examination of physical geography texts published over the past forty years showed some discussion of Thornthwaite's contributions (usually to the water budget or the climatic moisture index) in 33 of 55 books. This is truly a remarkable number of references for a geographer who has been dead for more than thirty years.

George Carter, chair of the geography department in which Thornthwaite taught during the late 1940s and early 1950s, summed up his impact:

> Warren Thornthwaite was an unforgettable man, both as a person and as a scholar with immense creativity combined with an intensely practical bent for solving problems. Such people come at long intervals and are rarely appreciated. Warren was fortunate in that his enormous contributions were recognized and that at Seabrook he had a nearly perfect laboratory to demonstrate the worth of his thinking.

Thornthwaite changed the course of climatology in the twentieth century by stressing that climatology was a rigorous discipline, defined by physical laws, and not merely the computing of average values of temperature and precipitation. He emphasized that the study of climatology was not limited to the realm of the geographer but belonged also to the botanist, the hydrologist and the engineer. He advocated the study of microclimates, or topoclimatology as he called it. As president of the Association of American Geographers, he challenged geographers to become more scientific and to use mathematical and physical principles in their research. He was a founder of the field of applied climatology. His water budget model, based on his novel concept of potential evapotranspiration, has proven to be useful to scores of scientists within and outside the realm of climatology. It has been used by hydrologists throughout the world as a means of quantifying the hydrologic cycle. It has seen increasing use in recent years as part of the burgeoning literature on the impact of possible climate change on water resources.

Warren Thornthwaite was a practical man, and he would be pleased to know that his ideas are helping modern society to solve practical problems. He possessed that spark of genius, that ability to postulate new ideas and to solve complex problems, that is rarely encountered. He clearly left his mark on climatology, on geography, and on the entire realm of science.

Bibliography and Sources

1. OBITUARIES AND ARCHIVAL SOURCES

Hare, F.K., 'Charles Warren Thornthwaite, 1899–1963', *Geographical Review*, Vol. 53, No. 4 (1963), 595–7.

Leighly, J., 'Charles Warren Thornthwaite, March 7, 1899–June 11, 1963', *Annals of the Association of American Geographers*, Vol. 54 (1964), 615–21.

Leighly, J., 'Drifting into geography in the twenties', *Annals of the Association of American Geographers*, Vol. 69, No. 1 (1979), 4–15.

Marcus, M.G., 'Coming full circle: physical geography in the twentieth century', *Annals of the Association of American Geographers*, Vol. 69 (1979), 521–32.

Mather, J.R. (ed.), Thornthwaite Memorial Volume I, Volume II', *Publications in Climatology*, Vol. 25, nos 2, 3 (1972).

Mather, J.R. and Sanderson, M., *The Genius of C. Warren Thornthwaite: Climatologist–Geographer*, University of Oklahoma Press, Norman OK (1996).

Slentz, F., Typed notes for Central Michigan University, Development Office (for Alumni Banquet, 15 October 1988).

Correspondence and literature sources on file in the C.W. Thornthwaite Laboratory of Climatology, Centerton, NJ.

Personal recollections from Marie Sanderson, J.R. Mather, F.K. Hare, US National Archives, Fred Thornthwaite, the Thornthwaite daughters (Elizabeth, Dorothy, Sally), Arnold Court, G.F. Carter, The University of California (Berkeley) Library and the University of Oklahoma Library.

2. SELECT BIBLIOGRAPHY OF WORKS BY C.W. THORNTHWAITE

1929 'The Polar Front in the interpretation and prediction of Oklahoma weather', *Proceedings of the Oklahoma Academy of Science*, Vol. 9, 93–9.

1931 'The climates of North America according to a new classification', *Geographical Review*, Vol. 21, 633–55.

1933 'The climates of the earth', *Geographical Review*, Vol. 23, 433–40.

1934 (with H.I. Slentz and C. Goodrich), *Internal Migration in the United States*, University of Pennsylvania Press, Philadelphia.

1936 'The Great Plains', in *Migration and Economic Opportunity: The Report of the Study of Population Redistribution*, University of Pennsylvania Press, Philadelphia, 202–50.

1937 'The hydrologic cycle re-examined', *Soil Conservation*, Vol. 3, 85–91.

1938 (with B. Holzman and D.I. Blumenstock) 'Climatic research in the soil conservation service', *Monthly Weather Review*, Vol. 66, 351–68.

1939 (with B. Holzman) 'The determination of evaporation from land and water surfaces', *Monthly Weather Review*, Vol. 67, 4–11.

1939 'The role of evaporation in the hydrologic cycle', *Transactions of the American Geophysical Union*, Vol. 20, 680–6.

1940 'Atmospheric moisture in relation to ecological problems', *Ecology*, Vol. 21, 17–28.

1940 'The relation of geography to human ecology', *Ecological Monographs*, Vol. 19, 343–8.

1941 *Atlas of Climatic Types in the United States 1900–1939*, US Department of Agriculture Miscellaneous Publication 421, Government Printing Office, Washington DC.

1941 'Climate and settlement in the Great Plains', in *Climate and Man*, US Department of Agriculture 1941 Yearbook, Government Printing Office, Washington DC, 177–87.

1942 (with C.F. Stewart Sharpe and E.F. Dosch) *Climate and Accelerated Erosion in the Arid and Semiarid Southwest, With Special Reference to the Polacca Wash Drainage Basin*, US Department of Agriculture Technical Bulletin 808, Government Printing Office, Washington DC.

1942 (with B. Holzman) *Measurement of Evaporation from Land and Water Surfaces*, US Department of Agriculture Technical Bulletin 917, Government Printing Office, Washington, DC.

1943 'Atmospheric turbulence and the measurement of evaporation', in

Proceedings of the Second Hydraulic Conference, 1–4 June 1942, College of Engineering, State University of Iowa, Studies in Engineering, Bulletin 27, Ames, Iowa, 280–8.

1943 'Problems in the classification of climates', *Geographical Review*, Vol. 33, 233–55.

1943 (with John Leighly) 'Status and prospects of climatology', *Scientific Monthly*, Vol. 57, 457–65.

1944 (with H.G. Wilm) 'Report of the Committee on Transpiration and Evaporation, 1943–44', *Transactions of the American Geophysical Union*, Vol. 26, 86–93.

1945 (with M.G. Wilm) 'Report of the Committee on Evaporation and Transpiration, 1944–45', *Transactions of the American Geophysical Union*, Vol. 26, 292–7.

1946 'The moisture factor in climate', *Transactions of the American Geophysical Union*, Vol. 27, 41–8.

1947 'Climate and moisture conservation', *Annals of the Association of American Geographers*, Vol. 37, 87–100.

1948 'An approach toward a rational classification of climate', *Geographical Review*, Vol. 38, 55–94.

1949 'Climate and soil moisture in the tropics', *Geographical Review*, Vol. 39, 498–501.

1951 'Agricultural climatology at Seabrook Farms', *Weatherwise*, Vol. 4, 27–30.

1951 (with J.R. Mather) 'The role of evapotranspiration in climate', *Archiv für Meteorologie, Geophysik und Bioklimatologie*, series B, Vol. 3, 16–39.

1951 'The water balance in tropical climates', *Bulletin of the American Meteorological Society*, Vol. 32, 166–73.

1952 'A charter for climatology', *World Meteorological Organization*, Vol. 2, 40–6.

1952 'Evapotranspiration in the hydrologic cycle', in *The Physical Basis of Water Supply and Its Principal Uses, the Physical and Economic Foundation of Natural Resources*, Interior and Insular Affairs Committee, House of Representatives, US Congress, Government Printing Office, Washington DC, 25–35.

1952 'Grassland climates', *Publications in Climatology*, Vol. 5, No. 6, 1–14.

1952 'Klima-Kalender und Pflanzenwachstum', *Die Umschau*, Vol. 53, 522–4.

1952 'Operations research in agriculture', *Journal of the Operations Research Society of America*, Vol. 1, 33–8.

1953 'The place of supplemental irrigation in post-war planning', *Publications in Climatology*, Vol. 6, 9–29.

1953 'The water balance in arid and semiarid climates', in *Desert Research*, Proceedings of the International Symposium held in Jerusalem, 7–14 May 1952, Research Council of Israel, Special Publication 2, Jerusalem, 112–35.

1954 (with J.R. Mather) 'Climate in relation to crops', *Meteorological Monographs*, Vol. No. 8, 1–10.

1954 'Topoclimatology', in *Proceedings of the Toronto Meteorological Conference 1953*, Royal Meteorological Society, London.

1955 (with F.K. Hare 'Climate classification in forestry, *Unasylva*, Vol. 9, 51–9.

1955 (with J.R. Mather) 'The water balance', *Publications in Climatology*, Vol. 8, 1–104.

1955 (with J.R. Mather) 'The water budget and its use in irrigation', in *Water: The Yearbook of Agriculture 1955*, US Department of Agriculture, Washington DC, 346–58.

1956 'The air as a water absorbing medium', in Ruhland, W. (ed.) *Handbuch der Pflanzenphysiologie (Encyclopedia of Plant Physiology)*, Vol. 3, Springer Verlag, Berlin, 257–64.

1956 'Climatology in arid zone research', in White, G.F. (ed.) *The Future of Arid Lands*, American Association for the Advancement of Science Publication 43, AAAS, Washington, DC, 67–84.

1956 (with J.R. Mather) 'Microclimatic investigations at Point Barrow, Alaska', *Publications in Climatology*, Vol. 9, 1–51.

1956 'Modification of rural microclimates', in Thomas, W.L. (ed.) *Man's Role in Changing the Face of the Earth*, University of Chicago Press, Chicago, 567–83.

1956 (with J.R. Mather and D.B. Carter) 'Instructions and tables for computing potential evapotranspiration and the water balance', *Publications in Climatology*, Vol. 10, 181–311.

1957 'The task ahead in climatology', *World Meteorological Organization Bulletin*, Vol. 6, 2–7.

1958 (with C.E. Molineux, J.R. Mather and D.B. Carter) 'Estimating soil moisture and tractionability conditions for strategic planning', *Air Force Surveys in Geophysics*, Vol. 94.

1958 'Introduction to arid zone climatology', in *Climatology and Microclimatology: Proceedings of the Canberra Symposium*, UNESCO, Arid Zone Research, Vol. 11, 15–22.

1958 (with J.R. Mather) 'Microclimatic investigations at Point Barrow, Alaska. 1956–1958', *Publications in Climatology*, vol. 11, 59–239.

1959 (with J.R. Mather and D.B. Carter) 'Three water balance maps of Southwest Asia', *Publications in Climatology*, Vol. 12, 1–57.

1959 (with J.R. Mather) 'Investigations of the climatic and hydrologic factors affecting the redistribution of strontium-90 in the soil', *Publications in Climatology*, Vol. 12, 49–91.

1959 (with W.J. Superior, F.K. Hare and K.R. Ono), 'Measurement of vertical winds in typical terrain', *Publications in Climatology*, Vol. 12, 933–204.

1960 (with J.R. Mather and J.K. Nakamura) 'Movement of radiostrontium in soils', *Science*, Vol. 131, 1015–19.

1961 'American geographers: a critical evaluation', *The Professional Geographer*, Vol. 13, 10–12.

1961 (with W.J. Superior, J.R. Mather and F.K. Hare) 'The measurement of vertical winds and momentum flux', *Publications in Climatology*, Vol. 14, 1–89.

1961 'The task ahead', *Annals of the Association of American Geographers*, vol. 51, 345–56.

1961 (with W.J. Superior and J.R. Mather) 'Vertical winds near the ground at Centerton, NJ', *Publications in Climatology*, Vol. 14, 91–244.

1962 'The geographer's role in climatology', in Leidlmair, A. (ed.) *Hermann von Wissmann–Festschrift*, Geographisches Institut der Universität Tübingen, Tübingen, 81–8.

John Russ Mather is Professor Emeritus, Department of Geography, University of Delaware, Newark, Delaware, 19716.

Marie Sanderson is Adjunct Professor, Department of Geography, University of Waterloo, Waterloo, Ont., N2L 3G1, Canada.

Chronology

1899 Born 7 March at Pinconning, Michigan

1918 Matriculated, Central Michigan Normal School

1922 BA Central Michigan Normal School

1922 Taught high school science, Owosso, Michigan

1924 Entered University of California at Berkeley to study for PhD (geography)

1925 Summer employment, Kentucky Geological Survey, Frankfort, Kentucky

1927–34 Assistant professor, University of Oklahoma, Norman, Oklahoma

1934 Left Oklahoma, moved to Washington to work on Great Plains study

1935 Joined Soil Conservation Service as director of Climatic and Physiographic Research Section

1939 Completed basic research study providing a year's record of daily evapotranspiration from a grassed field near Washington

1944 Consulting on irrigation of sugar cane with government of Mexico

1946 Leave of absence from Soil Conservation Service to begin private consulting on irrigation at Seabrook Farms, NJ

1947 Established Johns Hopkins University Laboratory of Climatology at Seabrook Farms as institute for climatic research

1947	Professor of Climatology and Geography, Johns Hopkins University
1948	Published seminal paper on 'An approach toward a rational classification of climate'
1951	Elected first president, Commission for Climatology, World Meteorological Organization
1952	Established C.W. Thornthwaite Associates to handle growing consulting activities in agriculture, industrial waste water disposal, climatological pollution from nuclear reactors
1953	Re-elected president of the Commission for Climatology, World Meteorological Organization, for a second term
1954	Transferred the Laboratory of Climatology from Johns Hopkins to Drexel Institute of Technology
1955	Chaired working group of Commission for Climatology that outlined specifications for world climatic atlas project
1956–8	Directed two-year research study on microclimatic conditions at Pt Barrow, Alaska
1957–8	World Meteorological Organization and UNESCO sponsored trips to Middle East, India and Australia, as well as around the world for meetings and lectures
1959	Received Cullum Geographical Medal of the American Geographical Society
1960	Elected Honorary President of the Association of American Geographers
1963	Received honorary doctorate from Central Michigan University
1963	Died 11 June in Elmer, NJ

Index

The index is divided into two parts:

1. A general index, including personal names, organizations, conferences, societies, and geographical concepts, theories and research.

2. A cumulative list of biobibliographies which includes all the geographers listed in volumes 1–17 inclusive.

1. GENERAL INDEX

2. CUMULATIVE LIST OF BIOBIBLIOGRAPHIES